The Social Effects of Global Trade

The Social Effects of Global Trade

Quantifying Impacts Using Multi-Regional Input-Output Analysis

edited by

Joy Murray
Arunima Malik
Arne Geschke

PAN STANFORD PUBLISHING

Published by

Pan Stanford Publishing Pte. Ltd.
Penthouse Level, Suntec Tower 3
8 Temasek Boulevard
Singapore 038988

Email: editorial@panstanford.com
Web: www.panstanford.com

British Library Cataloguing-in-Publication Data
A catalogue record for this book is available from the British Library.

The Social Effects of Global Trade

ISBN 978-981-4774-55-0 (Hardcover)
ISBN 978-020-3728-71-0 (eBook)

Cover design based on artwork by
Helen Winthorpe Kendrick
Textile Artist
Middleton St George
United Kingdom
The artwork is called 'Red Hot Chilli Pepper' and is made from a mixture of fabrics, mainly cotton but some man-made fabrics as well.

Printed in the USA

Contents

Foreword

Global trade has both liberated and put new constraints on the world and society. It allows a worker in China to help someone in Australia when that person's computer malfunctions. Global trade also provides a connection between a farmer in Peru and a consumer in Britain. The benefits that we can see can be massive. No longer are individuals limited by the market within which they are located—they can also reach the world quickly and relatively cheaply. With the opening of markets, so has the opportunity opened up for sharing knowledge, information and wealth. Technological leaps have been made more possible through globalisation and global trade. For example, countries in Africa, where citizens previously had limited access to banks and financial capital, have skipped over the brick-and-mortar stage of physical banks to go straight to online banking, requiring only a smart phone and Internet access. They are able to send money to family and friends, pay for goods and services, apply for a loan to start a new business and take payments for trade from around the globe.

However, there is also a darker side to international trade. Whereas once we understood how our food was grown, because we either did it ourselves or traded with someone in our local community, we are now often distanced from production. We don't really understand the production practices, the environmental impacts or how well people who are producing our food are being treated. Many consumer goods are a sum of their parts, and their parts could have been flown or shipped from anywhere in the world to central points for different stages of production and then redistributed around the world to waiting markets. Tracking and tracing trade can be hugely complex, requiring the use of both big data and small, local data sets. The complexity of global trade has increased significantly, with supply chains lengthening. As supply chains lengthen, they often become more complex and intertwined, looking more like supply webs than supply chains. There is a tendency for manual labour to be focused in certain countries and regions where labour is less expensive than developed countries and often regulations of industry are lower. These countries and industries become net exporters of labour, where their own workers don't benefit from global trade either financially, for their health or well-being, or for social justice. Alternatively, the labour force will travel to where the work is, creating an often stateless group of people working in some of the poorest conditions in developing countries. Analysis shows a growth in consumption of goods in developed nations since the mid-1990s, and yet Oxfam's latest report into wealth distribution indicates that eight men hold wealth equivalent to the poorest half of the world's population.[a] When I first started seriously researching the social impacts of global trade, the world's richest people could all fit into a double-decker bus. Now, in a relatively short space of time, they could fit into a minivan.

We rely on news services to tell us what the problem is, although in the post-truth world, with alternative facts thrown in for good measure, it is often difficult to understand what genuinely is the issue and how we, as consumers, academics, civil society, businesses or governments, can influence the problem being flagged to us. *The Social Effects of Global Trade: Quantifying Impacts Using Multi-Regional Input-Output Analysis* can help us address some of the fact deficit around the social impacts of global trade. Written in collaboration with leading authors from across the world, this book uses different forms of analysis to identify how labour, welfare and wealth flow with trade. Input-output analysis, life cycle assessment and social footprinting are all put to excellent use to help develop our understanding of the impacts of global trade beyond the headlines. Topics such as how trade impacts well-being, how trade can influence poverty reduction or inequality and the traceability of supply chains are all addressed here.

From an industry perspective, getting an accurate understanding of the social and environmental impacts of global supply chains for particular commodities is crucial. One cannot begin to resolve the complex issues of human rights, welfare, safety and inequality arising out of global supply chains without any understanding of the embodiment of labour in trade. Working in some of today's highest-profile industries for modern slavery, I know that nothing beats ground knowledge of the supply chain from sourcing through to production to distribution. However, getting that in-depth knowledge for a multitude of supply chains is rarely possible or

[a]https://www.oxfamamerica.org/explore/research-publications/an-economy-for-the-99-percent/

practical. This book helps us understand the intricate links between society and trade and how analysis can lead to solutions.

Darian McBain

Global Director of Sustainable Development
Thai Union
February 2017

Foreword

In the village where I grew up, the social cost of living consisted largely of the loss of one's own reputation, or that of one's family, as a consequence of having done or said something at odds with the prevalent social norms. The food that we ate was mostly grown in our own garden or on regional farms and vineyards, paper came from the local factory and brown coal came from a nearby mine. For me, as a child there existed no such thing as a zucchini or a Japanese car, fair trade was an unknown concept and child labour was something everyone thought was normal, as far as I could see.

This situation has changed drastically. While in 1970 world trade amounted to about 10% of the world's GDP, this figure now stands at about 30% [1]. Similarly, rapid development of information and communication technology means that we are now informed about the social predicaments of people living in the most remote corners of the world. We know that just eight people own the same amount as half the world population [2], 10% of the world's population lives on less than $1.90 a day [3] and more than two million people die each year as a result of occupational accidents or work-related diseases [4].

The contributions to this book demonstrate comprehensively how individual consumption by citizens of affluent societies is entwined with social impacts in less wealthy countries, how these impacts are often out of the consumers' sight (and mind?) and how international trade plays a decisive role in facilitating these impacts.

Increased knowledge entails increased responsibility. In his essay on famine, affluence and morality, Singer [5] writes on Bangladesh:

> I shall argue that the way people in relative affluent countries react to a situation like that in Bengal cannot be justified; indeed, the whole way we look at moral issues—our moral conceptual scheme—needs to be altered, and with it, the way of life that has come to be taken for granted in our society. [. . .] If it is in our power to prevent something bad from happening, without thereby sacrificing anything of comparable moral importance, we ought, morally, to do it.

Interestingly, in interpreting the moral imperative thus formulated, Singer picks up on the aspect of proximity and distance:

For the principle takes, firstly, no account of proximity or distance. It makes no moral difference whether the person I can help is a neighbor's child ten yards from me or a Bengali whose name I shall never know, ten thousand miles away.

In a way, the research presented in this book quantifies this proposition. Through a complex network of international supply chains, our consumptive activities become implicated with individual illnesses, injustices and unmet needs far away from us. Through this book, we are now able to find out how our own lifestyles are linked with worrisome social conditions elsewhere. This knowledge, according to Singer, extends our moral responsibilities.

When Singer wrote his essay I was a young child and, as such, would not have been able to appreciate his point of view. Most probably, there also existed very limited if not no information for the villagers to find out where their oranges, coffee or clothes came from, had they been interested in knowing. Having read this book today, I can no longer pretend ignorance when visiting the next shopping complex.

Manfred Lenzen
Sydney
February 2017

References

1. UNSD. (2017) *National Accounts Main Aggregates Database* (United Nations Statistics Division, New York), http://unstats.un.org/unsd/snaama/Introduction.asp.
2. Oxfam International. (2017) *Just 8 Men Own Same Wealth as Half the World*, http://www.oxfam.org/en/pressroom/pressreleases/2017-01-16/just-8-men-own-same-wealth-half-world.
3. World Bank. (2017) *Poverty & Equity Data*, http://povertydata.worldbank.org/poverty/home/.
4. International Labour Organisation. (2017) *Safety and Health at Work*, http://www.ilo.org/global/topics/safety-and-health-at-work/lang–de/index.htm.
5. Singer, P. (1972) Famine, affluence, and morality, *Philosophy and Public Affairs*, **1**, 229–243.

Preface

International Trade and Age-Old Dilemmas

On Saturday, 23 June 1888, a particularly cold summer's day in London, Annie Besant [1] wrote in the *Link*:

> But who cares for the fate of these white wage slaves? Born in slums, driven to work while still children, undersized because underfed, oppressed because helpless, flung aside as soon as worked out, who cares if they die or go on the streets, provided only that the Bryant and May shareholders get their 23 per cent, and Mr. Theodore Bryant can erect statues and buy parks?

She was talking about the plight of women and girls who made matches. The phosphorous used to make the matches caused phosphorus necrosis of the jaw, known as *phossy jaw*, which was incredibly painful, disfigured the face, and in 20% of cases led to the death of the sufferer. Yellow phosphorus was banned in Sweden and the United States, but the British government argued that banning its use was a restraint on free trade.

Restraints on free trade are still a major bone of contention. Can one country tell another what its social and environmental regulations should be? Under World Trade Organization (WTO) agreements, can one country ban imports from another on grounds of *methods of production* as well as *quality of product*? In 1991 Mexico complained under the General Agreement on Tariffs and Trade (GATT) dispute settlement procedure that a US ban on imports of its tuna was unlawful. The United States had banned imports of Mexican tuna because the United States believed Mexican fishing methods had not met US dolphin protection standards. The panel reported to GATT members in 1991 that 'the US could not embargo imports of tuna products from Mexico simply because Mexican regulations on *the way tuna was produced* did not satisfy US regulations'.[a] The reasoning behind the ruling was that one country could not impose

[a]http://www.wto.org/english/tratop_e/envir_e/edis04_e.htm (accessed 14 September 2014); italics in the original.

its environmental, health, and social standards on another country even, as in this case, 'to protect animal health or exhaustible natural resources'. Twenty years on, the dispute remained unresolved, although by then it was mainly focused around what restrictions, if any, could be applied to the use of dolphin-safe labelling of imported Mexican tuna.[b]

Three years after the United States banned tuna imports, Mexico entered into the North American Free Trade Agreement (NAFTA) with the United States and Canada, which, according to the International Monetary Fund (IMF),[c] saw an average annual real per capita gross domestic product (GDP) growth in Mexico roughly half that of the rest of Latin America over the following 20 years. During that time Mexico's poverty rate fell only 8 percentage points compared with 20 for the rest of the region. Mexican corn could not compete with the highly subsidized US corn, putting around 2 million Mexican farmers and their families out of work and contributing to the flow of poor Mexicans into the United States looking for work. And while Mexico's official Gini coefficient suggested some slight improvement in overall equality, the combined net worth of 15 Mexican billionaires reported in the Forbes rich list reached $150 billion.

At the same time another dispute was underway, this time between the European Union (EU), wanting to ban import of all seal products because of cruelty in the slaughter and skinning of seals, and exporters Norway and Canada, who were putting the argument to the WTO that banning their seal imports was discriminatory.

Whatever we think of today's free trade dilemmas, inter-country trade has been with us for centuries and has never been 'free'. There has always been a price to pay. However, who has paid that price has differed over time, place, and industry, as well as where you are in the supply chain. Around the same time as Annie Besant was writing in the *Link*, William Lever, later to merge with the Dutch company Unie to become Unilever, began building a model village, Port Sunlight, for his workers in England. His workers' cottages had running water and indoor bathrooms. His village had schools, a swimming pool, a library and a theatre. But while his workers in England were faring

[b]http://www.wto.org/english/tratop_e/dispu_e/cases_e/ds381_e.htm (accessed 14 September 2014).

[c]Weisbrot, M., Lefebvre, S. and Sammut, J. (2014). *Did NAFTA Help Mexico? An Assessment after 20 Years*, http://www.cepr.net/documents/nafta-20-years-2014-02.pdf.

well, workers on his palm oil plantations in the Belgian Congo were toiling in horrific conditions, treated little better than slaves.

It seems it was a similar story for George and Richard Cadbury, who, whether by ignorance or by oversight, relied for several years on cacao from plantations in Sao Tome and Principe, which depended on slave labour [4]. Meanwhile, at home, motivated by a concern for working conditions, health and safety, education and training, and affordable housing, the Cadbury brothers were revolutionising industrial relations on a national scale. Their personal commitment took place alongside a parliamentary debate on child labour, education and working hours, all of which contributed to lasting improvements for workers in England.

Similar debates were also happening elsewhere. In their book *Ethics*, published in the United States in 1909, John Dewey and James Tufts [3] discussed the ethics of the economic life. They considered the moral costs of producing cheap goods.

> The price of goods or service is not measurable solely in terms of other commodities or service; the price of an article is also . . . what it costs in terms of human life. . . . The introduction of machinery for spinning and weaving cotton cheapened cotton cloth, but the child labor . . . involving disease, physical stunting, ignorance, and frequently premature exhaustion or death, made the product too expensive to be tolerated. At least, it was at last recognized as too expensive in England; apparently the calculation has to be made over again in every community where a new system of child labor is introduced.

They could have added 'in every age', too. It seems that child labour is still with us, along with poor working conditions, exploitation of women and inequality of income. So also, at the other end of the spectrum, there are global companies committed to addressing such inequities, companies like today's Unilever, partnering with non-government organisation (NGO) Solidaridad and being committed to improving the lives of 1 million workers in its supply chain[d] [2].

---

[d]In November 2016 Amnesty International [2] reported that it was 'highly likely that Unilever . . . [was] sourcing palm oil from refineries where palm oil has been directly supplied or, at the very least, been mixed with palm oil produced on plantations where there are severe labour rights abuses'. Such a finding points to the difficulties encountered in trying to trace supply chains and the importance of the work of researchers who develop techniques for doing just that and those who use those techniques to meticulously examine supply chains in order to uncover social consequences of trade.

Consumers, too, are contributing to public discourse on the ethics of trade and pressuring companies for greater transparency about where, by whom and under what conditions their goods and services are produced. And for those who ignore the ethical argument there is the question of social unrest and its consequences. Almost a century after Dewey and Tufts wrote their book, philosopher Peter Singer [5] discussed the possible consequences of inequality in his book *One World: The Ethics of Globalization*:

> How well we come through the era of globalization (perhaps whether we come through it at all) will depend on how we respond ethically to the idea that we live in one world. For the rich nations not to take a global ethical viewpoint has long been seriously morally wrong. Now it is also, in the long term, a danger to their security.

At that time the UN's Department of Economic and Social Affairs would probably have agreed; in 2010 it released a paper 'The Inequality Predicament: Report on the World Social Situation, 2005'. The authors argued that inequality is a social justice issue as well as a threat to social, political and economic stability around the world. Addressing this inequality today is part of a continuous struggle; only the context changes as the passing of time makes yesterday's actors and events part of the context in which we act today.

Our book is a contribution towards understanding some of the threads that need to be untangled from the web of life, and examined, if we are to respond ethically to living in one world. We show how complex supply chains can be traced around the world and their social consequences exposed for good or ill. We follow commodities from producer to consumer, no matter how many countries they pass through in the production process or how complex that process is. Just as others have quantified the economic and environmental effects of doing business we quantify the social effects. For this we use input-output (IO) analysis. In illustrating the use of IO we hope to convince organisations of its value as a tool to examine the social effects of trade embodied in supply chains. We hope also that this book will be of use to students of economic, environmental and social sustainability in their quest for a cohesive story about global trade—its history, ethics, effects and extent. Moreover we hope this book provides an understanding of IO and its uses and provokes further study of the use of IO as a tool. To this end we collect together the work of academics in the forefront of using IO to analyse, account for and report on the social consequences of trade within and between countries.

Organisation of the Book

The book is divided into four parts. Part I, 'Introduction', sets the scene. It provides the origins, context and some of the implications of today's global trade network, as well as a human rights and social justice perspective on supply chain responsibility. Part II, 'Input-Output Analysis', describes the tools that are the focus of this book, tools that we offer as a powerful technique for putting numbers to the supply chain responsibility described in Part I. It describes methods and models for examining the social pillar of the triple bottom line, helping companies identify and quantify inequalities so that they can address them. Part III provides case studies—some good news stories as well as issues—illuminated by researchers committed to unpacking the intricacies of complex supply chains. Part IV concludes the book, providing some thoughts on what the future of social analysis, assessment and reporting might hold for business.

Part I: Introduction

In Chapter 1 Christopher Wright provides a broad historical context for our work. He traces the origins of trade within a capitalist society before arriving at the issues of trade in today's highly globalised world. He discusses the growing inequality and environmental degradation that threaten our future. He ends with a plea for detailed quantitative studies of the highly complex trade data collected by local, national and global organisations. He suggests that recent developments in IO analysis make it ideally suited to the job.

In Chapter 2 Christine Nolan provides a human rights and social justice perspective on regulation of global supply chains. She discusses the question of responsibility, questioning where in this globalised world of business responsibility begins and ends, who is responsible and how responsibility can be shared among all the myriad players in such complex supply chains. Theories of responsibility are presented, and the idea of legal accountability is discussed as a step towards regulation. Some examples of supply chain transparency legislation are given, illustrating how legal liability can be distributed up and down a supply chain and can become a mechanism for improving the human rights of vulnerable workers.

Part II: Input-Output Analysis

Chapter 3, written by Arne Geschke, provides an overview of how global social impacts can be identified and qualified. The methodology that is used is known as multi-regional input-output (MRIO) analysis, which is based on global accounting frameworks in combination with eligible social data. The author gives an overview of the concepts behind MRIO analysis and discusses which types of social data can be used for this approach. This chapter contains an introduction to the theoretical and methodological background that the case studies featured in this book are based on.

In Chapter 4, Catherine Benoît Norris, Gregory Norris, Yanyan Xiao and Joy Murray provide a means of tracking progress towards the Sustainable Development Goals (SDGs) launched by the UN using MRIO analysis combined with the Social Hotspots Database (SHDB). They take us step by step through the process of bringing together the Eora MRIO database and the SHDB and then provide a case study showcasing how governments and organisations might use this combined power to address human rights due diligence along supply chains.

Part III: Case Studies

In Chapter 5 Yanyan Xiao, from the University of Sydney, illustrates how non-additive social information can be manipulated into a form that can be used in social IO analysis. She uses national corruption indices and employment data by sector to estimate the corruption footprints of 189 nations. She finds that developed countries are net importers of corruption embodied in goods and services, while developing nations are net exporters of embodied corruption. Xiao's work reveals which commodities are the most highly implicated in the flow of corruption. Such information can be useful to corruption fighters everywhere.

In Chapter 6 Rachel Reyes, from the University of Sydney, uses a global MRIO trade database in conjunction with information on social welfare to account for the social welfare associated with consumption of goods and services. Her results uncover inequality in terms of welfare of workers in rich and poor nations. Furthermore, Reyes argues the need for a metric for measuring a country's success beyond merely relying on GDP.

Chapter 7, written by Ali Alsamawi and Yanyan Xiao, examines the flow of employment embodied in imports and exports in the Arab region. They identify flows between one Arab nation and another as well as those between Arab nations and the rest of the world. They define what he calls the *employment footprints of nations* as a measure that includes the number of workers working for the domestic market plus the number of workers embodied in its traded imports. Such footprint information assists a country in identifying how much employment it requires in addition to its own workforce in order to satisfy the nation's consumption demand. Thus it can be useful in designing appropriate employment policy.

Moana Simas and Richard Wood, from the Norwegian University of Science and Technology, Trondheim, contribute Chapter 8, in which they examine the distribution of labour and wages embodied in European consumption. The authors discuss the impact of buying products such as food, clothing and electronics on wages, employment and, in particular, vulnerable employment, embodied in the supply chains of these commodities.

In Chapter 9 Yafei Wang and Lixiao Xu use a comprehensive sub-national IO table of China to calculate inequality footprints of the country's provinces. Their results uncover top importers and exporters of inequality embodied in the consumption of goods and services. They assert that inequality assessments are useful for understanding disparity in terms of income and wages and can inform future policy-making.

In Chapter 10 Simon Mair, Angela Druckman and Tim Jackson present us with an in-depth study of the Western European clothing supply chain. Their analysis shows that while the Western European clothing supply chain could be considered 'fairer' because of globalisation, it can in no way be considered to be 'fair'. They find that workers in BRICS (Brazil, Russia, India, China and South Africa) countries are paid insufficient wages for a decent life.

Chapter 11 was written by Ruslan Lukach and José M. Reuda-Cantuche from the European Commission's Joint Research Centre, in Seville, Spain. They provide an overview of the main trends and patterns in income effects in global value chains driven by EU exports during 1995–2011. Their study allows examination of value added (income) per EU member state embodied in all EU exports. It shows the magnitude of interdependence within the EU and between the EU and the rest of the world's economy. The writers conclude that the single market is an important pillar of EU competitiveness in the global marketplace and contributes considerably to the well-being of the EU population.

In Chapter 12, Christina Prell, Klaus Hubacek, Laixiang Sun and Kuishuang Feng analyse effects of global trade on pollution and mortality. The authors interpret international trade as a network of interdependencies, in which each country's participation in international trade is represented by a level of integration within this network. Countries that are strongly interlaced in international trade are at the 'core' of the network, whereas countries with few economic ties outside their borders are situated in the network's 'periphery'. Using this classification, the authors then analyse whether core countries are able to outsource emissions from their home soil and whether this yields positive effects within the population. The authors find that for SO_2 emissions, the level of integration is positively associated with SO_2 emissions domestically, and even more so, within the international supply chain. Hence, 'core' countries are strong emitters at home and abroad. The authors further show that there is a strong negative correlation of a country's level of integration and the infant mortality rate—despite the higher domestic SO_2 emissions that were shown before. Hence, high emissions do not necessarily correlate to negative health risks, and the authors offer a number of explanations such as the pollution-wealth ratio to analyse their findings further.

Part IV: Social Effects of Global Trade in Standard Business Practice

Chapter 13, contributed by Graham Sinden, takes a step back from the case studies of the previous part and gives an overview of regulatory and non-regulatory initiatives concerning human rights issues along supply chains. While a number of states now have regulatory frameworks in place that specifically target human rights issues along supply chains outside their national territory, these regulations often have a specific focus (such as human trafficking and slave labour), and the term 'supply chain' is often lacking a clear definition. Non-regulative initiatives, mostly advocated by NGOs, often take a broader view and consider social issues along supply chains more broadly. Consequently, for businesses that are seeking to demonstrate their social responsibility, both regulatory and non-regulatory approaches are important and the interest in these is growing steadily. The author further discusses the challenges of quantifying social supply chain impacts and closes by remarking that social responsibility is already becoming a valuable non-financial performance measure for businesses and investors alike.

References

1. http://www.mernick.org.uk/thhol/thelink.html.
2. Amnesty International. (2016) *The Great Palm Oil Scandal: Labour Abuses Behind Big Brand Names* (Amnesty International, UK), p. 106, https://www.amnesty.org.uk/sites/default/files/the_great_palm_oil_scandal_lr.pdf.
3. Dewey, J. and Tufts, J. H. (1908) *Ethics* (Henry Hold, New York), https://archive.org/details/ethics02tuftgoog.p 489/90.
4. Satre, L. J. (2005). *Chocolate on Trial: Slavery, Politics, and the Ethics of Business* (Ohio University Press, USA).
5. Singer, P. (2002) *One World: The Ethics of Globalization* (Yale University Press, USA), p. 13.

Part I

INTRODUCTION

Chapter 1

Global Trade in an Era of Neoliberal Capitalism: Origins, Context and Implications

Christopher Wright
University of Sydney Business School, Sydney, NSW 2006, Australia
christopher.wright@sydney.edu.au

The exchange of goods and services has been central to the history of human development. However, interpretations of trade have varied from the belief in such exchange as a source of economic well-being and cultural enlightenment to a far more negative focus on the exploitation and violence that often accompanied such economic change. For instance, trade underpinned the global expansion of the great European empires of Spain, Portugal, France and Britain from the 1500s through the 1800s and assisted the flourishing of the European Renaissance and the culture and wealth that accompanied the birth of modern civilisation [1, 2]. From the European perspective, trade was often viewed through a romantic vision of heroic adventure and the triumph of man over nature. However, trade was also linked to the genocide and destruction of

The Social Effects of Global Trade
Edited by Joy Murray, Arunima Malik, and Arne Geschke

ISBN 978-981-4774-55-0 (Hardcover), 978-020-3728-71-0 (eBook)
www.panstanford.com

whole civilisations in the newly colonised worlds of the Americas, Africa and Australasia. Our conceptual understanding of trade has been similarly bifurcated—torn between the rarefied, functional thinking of economic theory in which *comparative advantage* and *gains from trade* are promoted [3, 4] and the more critical insights of sociologists and political economists studying the real-world consequences of trade in the developing and developed worlds [5, 6].

In this introductory chapter, I trace the origins and broader context of trade within capitalist society before focusing on the contemporary dynamics of trade in a globalised world. We are currently living in an era of unprecedented social and environmental crisis. Global trade and the neoliberal economic system that underpins it are driving our civilisation towards an abyss we may be unable to avoid. Growing income inequality and environmental threats such as climate change and biodiversity loss threaten our future. We badly need more detailed analysis and understanding of the true costs and benefits of global trade if we are to enact political strategies that respond to the existential challenges we will face this century. This is what this book seeks to do.

1.1 Origins: Trade and Global Capitalism

Trade has underpinned the development of human civilisation from its earliest days. From the exchange of flint and obsidian in prehistoric times, to the emergence of trade routes in the ancient civilisations of Egypt, Sumer, China, Greece and Rome, to the colonial expansion of the Portuguese, Spanish, French and British empires from the sixteenth to nineteenth centuries, trade has been a central driver of economic and social development. While some have claimed that profit-oriented entrepreneurship has been a feature of trade from its earliest days [7], others have critiqued such a view, arguing that a range of exchange systems operate beyond market relations embedded within social, cultural and religious institutions [8, 9].

During the Industrial Revolution in Britain in the eighteenth and nineteenth centuries, trade became central to theoretical understandings of economic prosperity. The classical economists

Adam Smith [4] and David Ricardo [3] developed what came to be known as the *free trade doctrine*, which argued that each country has a comparative advantage in particular exports based on its geographical position, access to resources and internal capabilities. Through specialisation in these areas, it was argued, each nation achieves gains from trade. Under the free trade view, the greatest economic advantage in terms of increased consumption and income accrued to nations that reduced national barriers to trade and maximised their particular areas of comparative advantage [10].

Despite the widely accepted view that free trade underpinned economic success, historical research suggests this is something of a misnomer. For instance, Chang [11] convincingly argues that Britain and the United States (and many other countries) relied heavily on aggressive tariff protection in their early industrialisation as a way of building new capacities within infant industries. For instance, despite a mythology of British laissez-faire industrial development, from 1720 to the 1840s the British government aggressively supported local manufacturers through cuts to import duties on raw materials, the abolition of export duties, subsidies for British-manufactured products and increasing duties on imported manufactured goods. Similarly, in the United States from 1816 to the end of the Second World War, high levels of tariff protection for manufactured imports combined with the natural protection of significant transport costs made US industries among the most protected in the world.

In both the United Kingdom and the United States, it was not until each country had reached a position of global industrial supremacy (in Britain's case from 1860 to 1930 and in the United States after 1945) that their governments then switched to aggressive advocacy for global free trade. As nineteenth-century German economist Friedrich List argued, advocacy for free trade by economically superior nations was simply a self-serving political strategy aimed at maintaining economic dominance [11]:

> It is a very common clever device that when anyone has attained the summit of greatness, he kicks away the ladder by which he has climbed up, in order to deprive others of the means of climbing up after him. In this lies the secret of the cosmopolitical doctrine of Adam Smith, and of the cosmopolitical tendencies of his great contemporary William Pitt, and of all his successors in the British Government administrations. (p. 4)

From this perspective, advocacy for free trade takes on an altogether different character, in which developed nations promote market liberalisation as a way of maintaining economic dominance, while opening up new markets for raw materials and cheap labour in newly industrialising economies. Indeed, the rise of the Asian 'tiger' economies of Japan, Korea, Taiwan and most recently China during the 1970s, 1980s and 1990s occurred along the very same path that earlier Western economies pursued: tariff protection, a strongly interventionist state, limitations on foreign investment, export subsidies, cartel arrangements, R&D support and public-private cooperation [11, 12].

1.2 Neoliberalism and the Washington Consensus

The benefits of free trade have often been asserted as an established economic truth, although they have also been subject to political opposition. In designing the post-war financial system, the United States and other Allied nations through the Bretton Woods agreement envisaged a world in which free trade would unleash economic growth and development. The creation of the General Agreement on Tariffs and Trade (GATT) in 1947 explicitly enunciated a free trade ethos in which markets were preferred to regulation, the theory of comparative advantage and gains from trade was re-emphasised and trade liberalisation was seen as the best way to generate economic growth and development for all [13]. While slow to be practically implemented by nation states during the post-war decades, over the last 30 years the pace of trade liberalisation has increased significantly, evident in the formation of the World Trade Organization (WTO) in 1995 following the 1986–1993 Uruguay round of GATT negotiations.

The growing acceptance of trade liberalisation coincided with the increasing globalisation of economic activities, the emergence of new markets and the growing power of multinational corporations (MNCs) [14, 15]. This process has been facilitated not only by new information technologies but more importantly by the rise to dominance of a philosophy of economic thinking in which market forces and the power of global corporations have been given free rein. The starting point for this process was the rise of neoconservative governments in the United Kingdom (with the election of Prime Minister Margaret Thatcher in 1979) and the United States (with the election of President Ronald Reagan in 1980), which marked the end of the post-Second World War Keynesian consensus and the emergence of what has come to be termed 'neoliberalism'. As Harvey [16] has argued:

> Neoliberalism is in the first instance a theory of political economic practices which proposes that human well-being can best be advanced by liberating individual entrepreneurial freedoms and private property rights, free markets and free trade. The role of the state is to create and preserve an institutional framework appropriate to such practices. (p. 2)

With the collapse of Eastern Bloc communist regimes in late 1989, free-market capitalism was seen as victorious in the battle of political ideas, what some free-market advocates claimed was 'the end of history' [17]. Neoliberal economic transformation now spread widely around the world, at the behest of Western governments, MNCs, free-market think tanks and management consultancies [18, 19]. In practice, neoliberalism has involved the deregulation of financial and labour markets, the privatisation of many government activities, and the withdrawal of the state from areas such as education, health, and social services [16, 20]. At the same time the state's role has increased in areas such as military spending, law enforcement and legal functions in order to bolster private property rights and the power of corporations in the new *market society* [21–23].

Moreover, this model of economic development has been globally disseminated by the International Monetary Fund (IMF) and the World Bank, as well as within various bilateral and multilateral trade agreements such as the North American Free Trade Agreement (NAFTA). For developing economies seeking financial support, a key part of any loan typically includes the neoliberal recipe of domestic economic reform, including privatisation and market liberalisation. As critics have noted, neoliberalism has become the hegemonic mode of discourse in contemporary capitalist society. The logic of the market has become ingrained within politics, the media and educational institutions and now shapes how the world is interpreted, lived and understood [16, 24].

1.3 The Free Trade Agenda: Implications for Social Well-Being

The globalisation of trade has been underpinned by the rise of the large MNC as the dominant agent of global economic activity [25]. Business corporations represent around 40% of the world's largest economic entities, their revenues dwarfing many national economies. This growing concentration of economic activity has also involved increasing control and coordination of global supply chains as MNCs seek out natural resources and cheap labour in the production of an ever-distending cornucopia of products and services. Information technology combined with the MNC's economic power has resulted in a compression of global time and space as economic value chains are streamlined to reduce cost and maximise efficiency [26]. A good example of this process is the US retail giant Walmart, which has become one of the world's largest corporations through a relentless pursuit of supply chain efficiency and cost reduction [27]. This is a global model of capitalist systemic rationalisation that has been mimicked by many other big-brand retailers and manufacturers, including companies such as Nike, Apple, IBM, Unilever, Nestle, General Electric and Coca-Cola, all seeking to maximise their competitive advantage in a global market of increasingly complex supply chain relationships [28, 29].

Critics of globalisation have argued that one of the key costs of this more general economic process has been the propensity for worker exploitation and abuse in developing economies as manufacturing is globally relocated to so-called cheap labour zones. This 'race to the bottom' has been demonstrated by a variety of examples, such as during the 1990s, when footwear company Nike was the subject of an on-going non-government organisation (NGO) campaign highlighting labour abuses, including low pay, long working hours, extreme work intensification, exposure to dangerous chemicals and physical abuse by supervisors. Nike responded by arguing it was not directly responsible for such activities as various levels of sub-contractors in countries such as Vietnam, Indonesia and China now manufactured its products. Nevertheless, the reputational damage that these campaigns had upon the Nike brand did force the company to bring in auditors to assess supply chain practices and introduce a code of conduct for all sub-contractors [30].

The Nike example has been played out many times subsequently as Western companies have outsourced production and service activities to newly industrialising economies that offer low-wage employees who lack even basic protections, let alone trade union representation. Large retailers have been a particular focus here, given their buying power within the supply chain and ability to ratchet down costs, often through sourcing cheap products produced under exploitative conditions. This has included the use of child labour in the production of many First World consumer goods, including the garment industry and food manufacturing [31]. More recently, revelations of extreme working conditions in China's booming computer manufacturing industry, including worker suicides, have highlighted the high cost of the developed world's everyday luxuries [32].

However, analysis of the social cost of globalised trade is divided on whether this has resulted in a general decline in working conditions. For instance, labour economists have argued that aggregate data suggest no wholesale reduction in working conditions and that, in fact, globalisation has resulted in overall improvements in working conditions for many in developing economies where prior conditions were worse (note, for instance, the propensity towards child and forced labour in many developing economy settings) [33, 34]. MNCs under pressure from NGO campaigns and aware of the reputational risk to their brand from supply chain scandals often introduce labour codes of conduct, which can be an advance on prior more exploitative working conditions, although their impact is variable [35].

Thus it is unclear whether globalised trade has resulted in an overall decline in wages and working conditions, and there is evidence of economic advance in developing economies. Foreign investment and the emergence of new industries in developing economies need not involve simple resource exploitation but may potentially involve more secure and higher-paid jobs than was previously the case. The outsourcing of call centre and information technology work to areas such as Bengaluru in India highlights how globalisation of trade can also generate more positive work and social outcomes [36]. Against this, however, are examples of increasing global demand for products such as mobile phones, which have generated whole new exploitative arrangements in developing economies, such as illegal coltan mining in Africa, which has helped fuel a violent civil war and resulted in countless fatalities [37, 38].

More generally, the globalisation of economic activities and neoliberalism appear to have promoted increasing inequality both in developed and in developing economies [39]. Indeed, in contrast to convergence theory, the gap between the richest and the poorest countries in real income per capita has widened dramatically [40]. As neoliberalism has promoted the retreat of the state from economic regulation and large corporations have gained greater power, so has the divide between rich and poor worsened [41].

1.4 A World Safe for Multinationals and Crossing 'Planetary Boundaries'

While the implications of the current free trade regime for social inequity and development are subject to debate, as the current climate crisis illustrates the environmental implications of globalised trade are far more explicit. While ecological modernists have argued that increasing economic development, technological innovation and environmental reform reduce pollution and environmental harm [42, 43], current environmental challenges such as climate change, ocean acidification and biodiversity decline highlight the fundamental conflict between conventional economic development and basic 'planetary boundaries' [44].

For instance, it is no coincidence that the dramatic escalation in global greenhouse gas emissions, deforestation and species extinction has occurred at precisely the same time that global economic development and trade has undergone its Great Acceleration since the 1960s [45]. Indeed, the worsening climate crisis highlights how neoliberal economic reform and the promotion of globalised trade are fundamentally in conflict with the need for radical decarbonisation of the world's energy system [46]. As Wright and Nyberg [47] have argued, corporate capitalism is now engaged in a process of 'creative self-destruction' in which the relentless pursuit of new global markets, the expansion of consumption and new forms of capital accumulation risk the demise of entire ecosystems and, indeed, a habitable climate for human civilisation. So global trade is driving the degradation of carbon sinks, such as forests and oceans, and the extinction of plant and animal species, as well as increasing the exploitation of fossil fuel reserves via new mega coal mines, tar sand processing, the fracking of shale and coal-seam gas and deep-water and Arctic oil drilling [48]. This occurs both directly through the global trade in energy commodities, such as coal, oil and gas, as well as indirectly through the ever-increasing global demand for raw materials and consumer products through increasingly complex global supply chains [49].

Despite the urgency of the unfolding climate crisis, what has been remarkable about the international response has been the wholesale rejection of any limitation upon economic growth, globalisation and free trade. For instance, while the international response to climate change has relied upon voluntary and non-binding agreements between nation states for greenhouse gas emission reductions, reinforced in the November 2015 Paris climate agreement, this stands in marked contrast to the draconian model the world's nations have applied to international trade, where agreements are highly prescriptive and enforced with punitive legal sanction [10].

This very different standard of international regulation has been most evident in the secret negotiations surrounding the Transpacific Partnership (TPP) and the Transatlantic Trade and Investment Partnership (TTIP). As the latest multilateral free trade deals being negotiated between the United States and various Pacific and Atlantic trading nations, the TPP and TTIP are notable not only for the breadth and extent of economic activities they seek to expose to the free trade paradigm but also in the establishment of a separate legal regime of enforcement that critics argue prioritises the rights of MNCs over nation states and their citizens [50]. In particular, provisions for *investor–state dispute settlements* (ISDSs) create a system of private arbitration within which global corporations can sue national governments for regulations that they believe will impair their profitability [51, 52]. Examples could include actions against governments passing public health laws to discourage smoking, regulating for minimum wages or improved worker health and safety or seeking to limit environmentally harmful resource extraction or pollution. As the ultimate expression of free trade for multinational capital, critics have argued such provisions would represent a form of post-democracy in which power moves from established political systems such as the judiciary and legislature within nation states to new private, globalised arenas where 'political elites do deals with corporate lobbies' [53].

Indeed, many examples of such privately arbitrated corporate legal action against nation states now exist under existing trade deals. For example, global tobacco giant Philip Morris has sued Australia and Uruguay under ISDS provisions in separate trade deals for introducing plain-packaging legislation in an effort to reduce smoking, French service company Veolia is suing the Egyptian government under ISDS clauses in a French–Egyptian free trade treaty for legislating for a small increase in the Egyptian minimum wage and pharmaceutical giant Eli Lilly is suing the Canadian government for US$100 million under ISDS provisions in the NAFTA because of the government's attempts to reduce the cost of medicines for its citizens [54].

As the latest assertions of a global free trade doctrine, critics argue the TPP and TTIP fundamentally conflict with global aspirations for improved social and environmental sustainability as enunciated in the UN's Sustainable Development Goals and the Paris climate agreement. The neoliberal belief in the wholesale benefits of trade and economic growth to solve urgent social and environmental crises places inordinate faith in the benevolence of MNCs and market forces, while winding back any capacity for citizens and democratic institutions to have much say over their future.

1.5 Global Trade for More Sustainable Outcomes?

In an era in which corporations and markets have increasing influence over the determination of national economic policy and, more fundamentally, threaten the future of a habitable climate on this planet, it is time to reconsider the hegemonic belief in the wholesale benefits of global trade based upon a neoliberal logic. While global trade has been central to human development both economically and socially, there is also a need for balance between the interests of capital and those of societies and environments. Our current era highlights the growing trade-offs between corporate profitability and shareholder value on the one hand and community and environmental well-being on the other. If we are to avert a crisis, we need to do more than place our faith in neoclassical economic theory.

While governments have become increasingly compromised within the corporatized economy [21, 23], there is a growing need for other social institutions to provide countervailing power in assessing the costs and benefits of global trade. Much of the critical work in holding corporations to account for their social and environmental impacts now falls on the shoulders of NGOs, trade unions, community organisations, activists and academics. Beyond qualitative analysis of the impacts of trade on local communities, there is also a critical need for quantitative analysis of the changing nature, extent and implications of an increasingly complex global trading system. Here, methods such as input-output analysis can provide valuable insight into not only who the key actors are within increasingly complex global supply chains but also the effects within donor and recipient economies in terms of a broad range of social, economic and environmental indicators. In this age of environmental and social crisis we need to go beyond outdated ideologies of free-market capitalism and the assumption that trade is a 'rising tide which lifts all boats'. We need to empirically test what works best for communities and environments from an increasingly interlinked and fast-paced global trading regime. It is high time we moved beyond the economic nostrums of eighteenth- and nineteenth-century economists and investigated the twenty-first-century world of globalised corporate capitalism through detailed analysis of complex data.

References

1. Braudel, F. (1982). *The Wheels of Commerce* (University of California Press, Berkeley).
2. Pagden, A. (2003). *Peoples and Empires: A Short History of European Migration, Exploration, and Conquest, from Greece to the Present* (Modern Library, New York).
3. Ricardo, D. ([1817] 1998). *Principles of Political Economy and Taxation* (Penguin, Harmondsworth).
4. Smith, A. ([1776] 1947). *The Wealth of Nations* (Dent, London).
5. Beckert, S. (2014). *Empire of Cotton: A Global History* (Alfred A. Knopf, New York).

6. Moses, A. D. (2008). *Empire, Colony, Genocide: Conquest, Occupation and Subaltern Resistance in World History* (Berghahn Books, New York).
7. Herskovits, M. (1952). *Economic Anthropology* (Knopf, New York).
8. Malinowski, B. (1948). *Magic, Science and Religion and Other Essays* (Free Press, Glencoe).
9. Polanyi, K. (1957). *The Great Transformation* (Beacon Press, Boston).
10. Dunkley, G. (2004). *Free Trade: Myth, Reality and Alternatives* (Zed Books, London).
11. Chang, H.-J. (2003). *Kicking Away the Ladder: Development Strategy in Historical Perspective* (Anthem Press, London).
12. Amsden, A. H. (1989). *Asia's Next Giant: South Korea and Late Industrialization* (Oxford University Press, New York).
13. Dunkley, G. (2000). *The Free Trade Adventure: The WTO, the Uruguay Round and Globalism* (Zed Books, London).
14. Beck, U. (2000). *What Is Globalization?* (Polity Press, Cambridge).
15. Guillén, M. F. (2001). Is globalization civilizing, destructive or feeble? A critique of five key debates in the social-science literature, *Annual Review of Sociology*, **27**, 235–260.
16. Harvey, D. (2005). *A Brief History of Neoliberalism* (Oxford University Press, Oxford).
17. Fukuyama, F. (1992). *The End of History and the Last Man* (Free Press, New York).
18. Kipping, M. and Wright, C. (2012). Consultants in context: global dominance, societal effect and the capitalist system, in *The Oxford Handbook of Management Consulting*, eds. Kipping, M. and Clark, T. (Oxford University Press, Oxford), pp. 165–185.
19. Mirowski, P. and Plehwe, D. (2009). *The Road from Mont Pelerin: The Making of the Neoliberal Thought Collective* (Harvard University Press, Boston).
20. Kelsey, J. (1995). *Economic Fundamentalism: The New Zealand Experiment; A World Model for Structural Adjustment?* (Pluto Press, London).
21. Barley, S. R. (2007). Corporations, democracy, and the public good, *Journal of Management Inquiry*, **16**(3), 201–215.
22. Giroux, H. A. (2004). *The Terror of Neoliberalism: Authoritarianism and the Eclipse of Democracy* (Paradigm, Boulder).

23. Sandel, M. J. (2012). *What Money Can't Buy: The Moral Limits of Markets* (Farrar, Straus and Giroux, New York).
24. Larner, W. (2006). Neoliberalism: policy, ideology, governmentality, in *International Political Economy and Poststructural Politics*, ed. Goede, M. (Palgrave Macmillan, London), pp. 199–218.
25. Bakan, J. (2004). *The Corporation: The Pathological Pursuit of Profit and Power* (Constable, London).
26. Harrison, B. (1997). *Lean and Mean: The Changing Landscape of Corporate Power in the Age of Flexibility* (Guilford Press, New York).
27. Lichtenstein, N. (2006). *Wal-Mart: The Face of Twenty-First Century Capitalism* (New Press, New York).
28. Altmann, N. and Deiß, M. (1998). Productivity by systemic rationalization: good work - bad work - no work?, *Economic and Industrial Democracy*, **19**(1), 137–159.
29. Dauvergne, P. and Lister, J. (2013). *Eco-Business: A Big-Brand Takeover of Sustainability* (MIT Press, Cambridge).
30. Spar, D. L. and Burns, J. (2002). Hitting the wall: Nike and international labor practices, *Harvard Business School Case*, 9-700(047).
31. Institute for Global Labour and Human Rights. (2015). *Dirty Toys Made in China: The Zhenyang Wanju Sweatshop Factory* (IGLHR, Pittsburgh).
32. Institute for Global Labour and Human Rights. (2014). *Exhaustion Has No Limit at Apple Supplier in China* (IGLHR, Pittsburgh).
33. Enderwick, P. (2006). *Globalization and Labor* (The New Global Society/Chelsea House, Philadelphia).
34. Flanagan, R. J. (2006). *Globalization and Labor Conditions* (Oxford University Press, New York).
35. Barrientos, S. (2008). Contract labour: the 'Achilles heel' of corporate codes in commercial value chains, *Development and Change*, **39**(6), 977–990.
36. Kuruvilla, S. and Ranganathan, A. (2008). Economic development strategies and macro- and micro-level human resource policies: the case of India's 'outsourcing' industry, *Industrial & Labor Relations Review*, **62**(1), 39–72.
37. Hayes, K. and Burge, R. (2003). *Coltan Mining in the Democratic Republic of Congo: How Tantalum-Using Industries Can Commit to the Reconstruction of the DRC* (Fauna & Flora International, Cambridge).
38. Moran, D., McBain, D., Kanemoto, K., Lenzen, M. and Geschke, A. (2015). Global supply chains of coltan, *Journal of Industrial Ecology*, **19**(3), 357–365.

39. Dreher, A. and Gaston, N. (2008). Has globalization increased inequality?, *Review of International Economics*, **16**(3), 516–536.

40. Anand, S. and Segal, P. (2015). The global distribution of income, in *Handbook of Income Distribution*, eds. Atkinson, A. B. and Bourguignon, F. (Elsevier, Amsterdam), pp. 937–979.

41. Piketty, T. (2014). *Capital in the Twenty-First Century* (Harvard University Press, Cambridge).

42. Hajer, M. (1995). *The Politics of Environmental Discourse: Ecological Modernization and the Policy Process* (Oxford University Press, Oxford).

43. Mol, A. P. J. (2002). Ecological modernization and the global economy, *Global Environmental Politics*, **2**(2), 92–115.

44. Rockström, J., Steffen, W., Noone, K., Persson, Å., Chapin, F. S., Lambin, E. F., Lenton, T. M., Scheffer, M., Folke, C., Schellnhuber, H. J., Nykvist, B., de Wit, C. A., Hughes, T., van der Leeuw, S., Rodhe, H., Sörlin, S., Snyder, P. K., Costanza, R., Svedin, U., Falkenmark, M., Karlberg, L., Corell, R. W., Fabry, V. J., Hansen, J., Walker, B., Liverman, D., Richardson, K., Crutzen, P. and Foley, J. A. (2009). A safe operating space for humanity, *Nature*, **461**(7263), 472–475.

45. Steffen, W., Crutzen, P. J. and McNeill, J. R. (2007). The anthropocene: are humans now overwhelming the great forces of nature?, *AMBIO: A Journal of the Human Environment*, **36**(8), 614–621.

46. Klein, N. (2014). *This Changes Everything: Capitalism vs. The Climate* (Simon & Schuster, New York).

47. Wright, C. and Nyberg, D. (2015). *Climate Change, Capitalism and Corporations: Processes of Creative Self-Destruction* (Cambridge University Press, Cambridge).

48. IEA. (2013). *Resources to Reserves 2013: Oil, Gas and Coal Technologies for the Energy Markets of the Future* (International Energy Agency, Paris).

49. Lenzen, M., Moran, D., Kanemoto, K., Foran, B., Lobefaro, L. and Geschke, A. (2012). International trade drives biodiversity threats in developing nations, *Nature*, **486**(7401), 109–112.

50. Tienhaara, K. (2006). What you don't know can hurt you: investor-state disputes and the protection of the environment in developing countries, *Global Environmental Politics*, **6**(4), 73–100.

51. Ranald, P. (2014). Expropriating public health policy: tobacco companies' use of international tribunals to sue governments over public health regulation, *The Journal of Australian Political Economy*, **73**, 76–102.

52. UNCTAD. (2015). *Investor-State Dispute Settlement: Review of Developments in 2014* (United Nations Conference on Trade and Development, Geneva).
53. Crouch, C. (2014). Democracy at a TTIP'ing point: seizing a slim chance to reassert democratic sovereignty in Europe, *Juncture*, **21**(3), 176–181.
54. West, M. (2015). Trade deals acronym really translates to 'we lose', *Sydney Morning Herald*, http://www.smh.com.au/business/comment-and-analysis/trade-deals-acronym-really-translates-to-we-lose-20150618-ghrqm8.html.

Chapter 2

Regulating Human Rights and Responsibilities in Global Supply Chains

Justine Nolan
University of New South Wales, Sydney, NSW 2052, Australia
Justine.nolan@unsw.edu.au

2.1 Introduction

Today's global supply chains link individual workers with large and small companies across national, political and cultural boundaries. These complex and vast supply chains are a litmus test for any corporate responsibility or business and human rights program that aims to enable respect for human rights.[a] The supply chain is an area of potential commercial (including reputational) risk for companies. Evolving acceptance of a company's responsibility to respect all human rights [1] is making it more difficult for businesses to disassociate themselves from human rights violations that inevitably arise in many global supply chains that the businesses may be directly

[a]Today, many companies, particularly consumer-facing global companies, have a dedicated corporate social responsibility, sustainability or business and human rights program. See www.business-humanrights.org (accessed 26 October 2016).

The Social Effects of Global Trade
Edited by Joy Murray, Arunima Malik, and Arne Geschke

ISBN 978-981-4774-55-0 (Hardcover), 978-020-3728-71-0 (eBook)
www.panstanford.com

or indirectly involved with. However, despite the rhetoric around the acceptance of a company's responsibility to respect human rights, ambiguity still remains as to whom such responsibility might be attributed to and whether such responsibility includes accountability for wrongdoings. Any desire or attempt to compartmentalise human rights and responsibilities to a particular company or within particular geographical boundaries is exacerbated by direct and indirect corporate linkages in global supply chains. Each company along that supply chain has a responsibility to respect human rights—but where does that responsibility begin and end? How might responsibility be shared, and is there a difference between responsibility and accountability?

2.2 Supply Chains

The supply chain has become an intrinsic part of the lead firm's operations[b] but generally comprises separate legal entities. Increasingly, companies do not generally own or operate the end factories in which their goods are produced and they may contract with hundreds, sometimes thousands, of different suppliers annually [2]. Some recent examples of corporate practice evidence an evolving willingness of (some) global companies to acknowledge the vital nexus that exists between the operations of their contractors and sub-contractors and the ultimate delivery of the companies' goods. Nike, for example, sources its products from over 655 factories, engaging more than 1 million workers across more than 40 countries.[c] Walmart, one of the world's largest and most powerful retailers, sells its goods in '11,500 stores under 72 banners in 28 countries and e-commerce sites in 11 countries each week. With fiscal year 2016 revenue of $482.1 billion, Walmart employs 2.2 million associates worldwide' [5]. Walmart acknowledges that it has 'thousands of businesses around the world that proudly call

[b]'Lead firm' refers to the firm at the top or apex of the supply chain. In this chapter, generally it refers to a multinational company that bears the primary responsibility for distributing the product to consumers.

[c]See Ref. [3]. Nike is one of the world's largest suppliers of athletic shoes and apparel and a major manufacturer of sports equipment, with the fiscal earning for 2016 in excess of US$32 billion [4].

themselves Walmart suppliers' [6]. Apple publicly acknowledges its top 200 suppliers but does not state how many more lie beneath them.

Acknowledgement of responsibility by lead firms for the human rights of workers in their supply chains, is not, however, the same as accepting that such lead firms could or should assume legal liability for wrongdoing by suppliers in their production chains. Requiring or 'requesting' suppliers to fund the reimbursement of recruitment fees paid by migrant workers (such as Apple has recently done)[d] or to pay increased wages in their factories is an acknowledgement of the universalisation of global responsibility for supply chain operations and suggests a degree of leverage on the part of the lead firms to influence the conduct of the suppliers. It does not, however, necessarily transform that responsibility into legal liability for lead firms.[e] Nevertheless, public statements about the breadth, depth and interdependent nature of supply chains by influential multinational

---

[d]Apple's 2009 *Supplier Responsibility Progress Report* highlighted a specific problem in its supply chain, namely the payment of recruitment fees by the migrant workers employed by its suppliers. The report noted that Apple's suppliers used multiple third-party labour agencies to source workers from other countries, and it was common practice to charge workers a recruitment fee. In response, Apple updated its *Supplier Code of Conduct* and issued a standard for the prevention of involuntary labour. In 2014 Apple announced that starting in 2015, no worker employed on an Apple line could be charged any recruitment fees. Since 2008 Apple has required its suppliers to reimburse what it regards as excessive recruitment fees. In 2014, Apple required its suppliers to reimburse US$3.96 million in excess fees to over 4500 foreign contractors, bringing the total reimbursements to US$20.96 million to over 30,000 foreign contract workers since it began addressing this issue in 2008 [7].

[e]However, legal developments in a very select few cases in Canada seem to signal an increased willingness by an Ontario court, for example, to impute the wrongs of a subsidiary operating in a foreign country to the parent corporation. See, for example, Choc v. Hudbay Minerals Inc., ONSC 1414 (2013). In Hudbay, a group of indigenous peoples from Guatemala sued Hudbay Minerals Inc., a Canadian corporation, and its Guatemalan subsidiary, Compañia Guatemalteca De Niquel (CGN), for alleged human rights abuses at a Guatemalan mining project owned through CGN. The plaintiffs brought their claims in Ontario, not Guatemala. Hudbay applied to strike the claims on the basis that the claims improperly relied on 'piercing the corporate veil' or ignoring the separate corporate personalities of Hudbay and CGN. Hudbay further argued that the plaintiffs were seeking to impose a supervisory liability on parent corporations over their foreign subsidiaries. Hudbay's applications were dismissed, and the action allowed to proceed against it and its Guatemalan subsidiary in an Ontario court. Also see Chevron Corp v. Yaiguaje, SCC 42 (2015).

companies are a welcome step and a far cry from the narrower perspective and denial-of-responsibility approach of companies just a few decades earlier.

2.3 Theories of Responsibility

Global supply chains diffuse responsibility for the maintenance of human rights among states, tens or hundreds of contractually discrete companies, civil society, trade unions, consumers and the workers themselves [3]. Can theory provide guidance for developing a model that may assist in preventing and redressing corporate violations of human rights in supply chains? A number of models have been espoused that are applicable to the attribution of corporate responsibility for human rights in supply chains.

A decade ago, Young [8] offered a *social connection model* of responsibility, which proposes 'that all agents who contribute by their actions to the structural processes that produce injustice have responsibilities to work to remedy these injustices' [8]. Young posits that global supply chains, such as those in the apparel industry, are examples of 'structural injustice' that occurs 'as a consequence of many individuals and institutions acting in pursuit of their particular goals and interests, within given institutional rules and norms' [8]. All those involved in the process are responsible in the sense of being part of the process, even though they may not have directed the process or intended the outcome (i.e., unsafe working conditions). Young argues that the purpose of the social connection model is forward-looking rather than about the attribution of blame and that responsibility may be differentiated according to the particular party's position within the structural process. The degree of responsibility can be reasoned according to four factors: (i) power (potential or actual power to influence other actors in the process), (ii) privilege (benefit obtained via the structural injustice, such as middle-class consumers of apparel), (iii) interest (those who have a strong interest in perpetuating the injustice or changing the process) and (iv) collective ability (the extent to which it is possible to work together). Young's model emphasises a notion of collective or shared responsibility and assumes that various actors who may be directly or even more tangentially involved in a supply chain will hold each other to account and share responsibility for overcoming the structural injustice. Young's model tends to focus largely on the role of the state, the lead firm and the suppliers. Other actors, such as trade unions, non-governmental organisations (NGOs), the workers and consumers, potentially also have a role to play in this model. The extent to which responsibility should be exercised individually is largely dependent on the position of the party in the process.

Five years later, the United Nations (UN) Special Representative on Business and Human Rights put forward the UN *Guiding Principles on Business and Human Rights*, which sets forward a 'complementary but separate' theory of responsibility that views states and businesses working together but fulfilling distinct roles and responsibilities towards human rights. States have a legal obligation to protect rights (grounded in international law), and companies have a responsibility to respect human rights (based on social expectations to do no harm). This theory reflects an element of Young's reasoning in that it assumes companies have some level of moral responsibility for adherence to human rights and some leverage to exert change. This moral responsibility might be distinguished from legal accountability, which is largely state-centric as violations by companies are principally a matter for the government of the country in which the company is operating [9, 10]. The guiding principles do incorporate an element of extended responsibility with respect to a corporate supply chain in that a company's responsibility to respect human rights applies not only to the company's 'own activities' but also to those impacts that are 'directly linked to their operations, products or services by their business relationships, even if they have not contributed to those impacts' [1].

More recently, in 2015, the World Economic Forum's [11] Global Agenda Council proposed a model of *shared responsibility* that seeks to overcome 'weak governance and unsustainable upstream business practices' [11] that are evident in the many places that global supply chains penetrate. The council is inclusive when considering the relevant parties that might share such shared responsibility and includes global companies, national-level suppliers, developed and developing governments, international financial and labour organisations, workers, trade unions, civil society, private philanthropy and development agencies in its analysis. The council argues that for this new paradigm to be successful the framework of shared responsibility must be inclusive of four critical components: (i) adopt industry-wide systemic approaches, (ii) increase visibility of human rights challenges in supply chains and delink visibility from financial responsibility, (iii) assess the real costs of addressing the issues and (iv) generate cooperative approaches based on an equitable sharing of responsibility among key stakeholders [11]. This model is also predicated on the need for collective action but proposes that responsibility be differentiated and argues that 'the special responsibilities of each stakeholder must be specified and agreed upon in order to pursue the group's collective goal' [11].

Each of these models emphasises responsibility to different degrees, and all propose some form of collective action to overcome systemic issues evident in supply chains resulting from what Young refers to as *structural injustice* or what the UN guiding principles call *governance gaps* or what the council acknowledges as *weak governance* and *unsustainable business practices*. The notion of collective action is most prominent in the social connection and shared responsibility models, but the UN guiding principles also acknowledge that both states and companies must work complementarily to improve respect for human rights.

Some problematic assumptions associated with these theories of responsibility have been identified by Frenkel, Mamic and Greene [12] in terms of their applicability to implementing change in supply chains. These include the assumptions that '(1) the actors in the collective can be easily identified and agreed upon; (2) allocation of roles, implementation processes and responsibilities can be collectively decided in a fair manner; . . . (3) sufficient information is available to make collective decisions about key governance issues that contribute to upholding or enhancing labour standards in the supply chain; . . . and (4) effective mutual accountability can be designed and applied' [12]. Particularly problematic is the assumption that accountability is built into these models. Responsibility does not necessarily equate to accountability, and these models embody an 'accountability deficit' that might be overcome with the support of state regulation to support such collective responsibility.

One might argue that responsibility and accountability are distinguishable terms. The concept of responsibility was seemingly deliberately used in the context of the guiding principles to distinguish a corporation's responsibility to respect human rights from a state's obligation to do so, the former being 'softer' and implying an undertaking that is something less than a legal obligation [13]. However, there is commonality between the terms, and indeed, the *Oxford English Dictionary* defines responsibility as both having control over someone or something and being accountable for something. Accountability seems to imply ownership of an outcome. Responsibility may be commonly shared (as these models theorise), but must accountability always be linked to a single entity and is there something more forceful in using the term 'accountable' rather than 'responsible'? The language of business and human rights, for example, has moved away from 'corporate responsibility' terminology and calls for advocacy are now more likely to be linked to accountability [14, 15]. Increasing demands for corporate human rights accountability are now enveloping the myriad of corporate actors and activities within a company's supply chain [2, 11, 16]. Being accountable implies that you are not only responsible for something but also ultimately answerable for your actions. Young, however, uses the term 'responsibility' interchangeably as used in the dual sense proposed by the dictionary meaning. The key distinction she argues is based not so much on terminology as on whether responsibility is shared and forward-looking versus responsibility that attributes blame or liability and is retrospective. The latter is a key part of what she calls the *liability model* [8] or what some might associate today more closely with corporate accountability. Young [8] describes the liability model as follows:

> One assigns responsibility to a particular agent [or entity] whose actions are connected to the circumstances for which responsibility is sought. . . . Where the actions were voluntary and were undertaken knowingly. . . . it is appropriate to blame the agent for the harmful outcomes.

Young then goes on to note that because of the causal connection required between the wrongdoing and the agent, the liability model is not appropriate to address structural social injustice. The World Economic Council, while not referencing Young's work or the liability model directly, also argues that a limitation of a compliance model 'is that it confines the definition of problems in the supply chains to those for which multinational companies will accept responsibility' [8]. The council suggests that if companies are asked to assume the sole responsibility for human rights issues in their supply chains, then they are more likely to define their supply chains narrowly.

I argue, instead, that both corporate responsibility and corporate accountability can be shared and that for the theories around shared responsibility to be viable, they must overcome their accountability deficit. That is, responsibility must also incorporate accountability (answerability for outcomes) but such accountability need not necessarily sit with only one actor in the chain. Indeed, legislative examples are emerging that demonstrate how legal accountability can be shared for human rights violations that take place in a supply chain (see the discussion later regarding the regulation of Australian homeworkers in the apparel industry). The danger of relying simply on notions of shared responsibility is that sometimes when everyone is responsible, no one is. The concept of collective action and shared responsibility is indeed actionable in the business and human rights context but will arguably be most effective when operating in the 'shadow of state regulation' [17]. The legislative initiatives mentioned briefly later demonstrate an emerging broad approach to legislatively incorporating notions of shared responsibility in a supply chain (e.g., a lead firm assuming responsibility for reporting on forced labour in its entire supply chain) and some incorporate mechanisms to share accountability when wrongdoings occur.

2.4 Regulating Responsibility

In 2015, the United Kingdom passed the Modern Slavery Act, which mandates transparency in supply chains. The UK legislation was modelled on, but is potentially broader than, the California Transparency in Supply Chains Act passed a few years earlier. This type of legislation is designed principally to produce information about steps taken by companies to conduct supply chain due diligence on specific human rights issues and to force companies that have not engaged in due diligence to say so publicly. While the American and British legislative developments mandate disclosure, they do not directly impose liability on lead firms for the downstream acts of other companies in their supply chain.

However, transparency legislation can also be crafted to expressly attach legal liability up and down a supply chain for particular wrongdoings occurring anywhere in that chain. This type of legislation emphasises the link between control and/or leverage and responsibility. If a lead firm at the top end of the supply chain can control the size, design, quantity and quality of a product and possesses potential leverage to influence the working conditions of those producing the goods, it is arguably then both fair and effective to align that power with legal accountability. One legislative example that has been implemented in Australia relatively recently combines responsibility for reporting with a notion of shared accountability for wrongdoing that may occur in the supply chain. The legislation regulates the working conditions of homeworkers in the Australian clothing industry and embodies a concept of shared accountability in that one or more actors in the chain might potentially be responsible for wrongdoing, even if the action was taken by a separate legal entity in the supply chain.

In the early 2000s, a number of state governments in Australia introduced supply chain regulation in response to a strong and sustained civil society campaign focused on safeguarding the rights of home-based workers in the clothing industry.[f] The Australian legislation set up a regulatory framework that requires the insertion of contractual tracking mechanisms in supplier contracts to follow production and mandate disclosure up and down the supply chain. From the retailers down, supply chain participant companies are obligated to establish whether homeworkers are being used in the production of their goods and suppliers at all levels are obligated to provide that information to parties further up the chain. The information is then provided by the retailer to a designated trade union so that the supply chain is transparent and able to be monitored. The legislation broadens the traditional approach to defining

[f]The FairWear campaign (http://fairwear.org.au/) was a significant force that led to the introduction of successive legislative amendments, including the Industrial Relations (Ethical Clothing Trade) Act 2001 (NSW), the 175B Workers Compensation Act 1987 (NSW), the Industrial Relations (Fair Work) Act 2005 (SA), the Outworkers (Improved Protection) Act 2003 (Vic) and the Industrial Relations and Other Acts Amendment Act 2005 (Qld). Also, amendments to the Fair Work Act 2009 (Cth) in 2012 largely adopted the NSW model (Part 6-4A). Also see generally Refs. [18, 19].

employment by deeming home-based workers or outworkers to be employees in order to defeat commercial arrangements that might otherwise undermine the employee status, for instance, by classifying workers as 'independent contractors'. In many cases, the immediate 'employer' may not be the controlling entity. Control may be shared by a number of key players in the supply chain or be wielded by a firm that sits outside the traditional employment relationship. The provisions also impose obligations on successive parties (other than the retail sector) in the contracting chain to ensure that homeworkers receive their lawful entitlements (rights of recovery). The legislation provides for a recovery mechanism whereby the workers are entitled to claim unpaid entitlements throughout the supply chain up to and including the principal manufacturer. Finally, the legislation is supplemented by a voluntary private mechanism administered by a multi-stakeholder initiative—Ethical Clothing Australia[g]—that accredits businesses operating in this specific sector and assists companies with the process of mapping their supply chains and verifying that all workers within these supply chains are receiving their legal entitlements. The legislation essentially views the disparate actors in the supply chain as part of a larger business enterprise and contemplates the notion that accountability for wrongdoings might be shared among separate companies in the chain.

2.5 Conclusion

The regulation of global supply chains creates opportunities and challenges in improving the practices of corporate entities with respect to human rights. Best-practice supply chain regulation may require lead firms to play a significant role in monitoring and reporting on their supply chains and may also transfer some level of legal accountability to those firms for the downstream activities of their suppliers. It can also incorporate a mix of public and private actors in the regulatory framework and ultimately should embody notions of both shared responsibility and shared accountability.

[g]http://www.ethicalclothingaustralia.org.au/

References

1. Human Rights Council. (2011). *Guiding Principles on Business and Human Rights: Implementing the United Nations 'Protect, Respect and Remedy' Framework*, Report of the Special Representative of the Secretary-General on the issue of human rights and transnational corporations and other business enterprises (A/HRC/17/31) March 21.
2. Locke, R. M. (2013). *The Promise and Limits of Private Power* (Cambridge University Press, New York).
3. Nike. *Frequently Asked Questions*, http://nikeinc.com/pages/frequently-asked-questions (current as of April 2016) (accessed October 26, 2016).
4. Nike. (2016). *Nike, Inc. Reports Fiscal 2016 Fourth Quarter and Full Year Results*, http://news.nike.com/news/nike-inc-reports-fiscal-2016-fourth-quarter-and-full-year-results.
5. Walmart. (n.d.). *About us*, http://corporate.walmart.com/our-story.
6. Walmart. (n.d.). *Suppliers*, http://corporate.walmart.com/suppliers.
7. Apple. (2014). *Supplier Responsibility 2014 Progress Report*, http://www.apple.com/supplier-responsibility/pdf/Apple_SR_2014_Progress_Report.pdf.
8. Young, M. (2006). Responsibility and global justice: a social connection model, *Social Philosophy and Policy*, **23**(1), 102–130.
9. Deva, S. and Bilchitz, D. (2013). When human rights 'responsibilities' become 'duties': the extra-territorial obligations of states that bind corporations, in *Human Rights Obligations of Business: Beyond the Corporate Responsibility to Respect?*, eds. Augenstein, D. and Kinley, D. (Cambridge University Press, Cambridge), pp. 271–294.
10. Nolan, J. (2013). The corporate responsibility to respect human rights: soft law or not law?, in *Human Rights Obligations of Business: Beyond the Corporate Responsibility to Respect?*, eds. Deva, S. and Bilchitz, D. (Cambridge University Press, Cambridge), pp. 138–161.
11. World Economic Forum. (2015). *Shared Responsibility: A New Paradigm for Supply Chains*, http://www3.weforum.org/docs/WEF_GAC_Supply_Chains_%20A_New_Paradigm_2015.pdf.
12. Frenkel, S., Mamic, I. and Greene, L. (2016). Towards shared responsibility: theory and practice in regulating labor standards in global supply chains, *Proceedings of the EGOS Conference, Naples*, pp. 11–12.

13. Ramasastry, A. (2015). Corporate social responsibility versus business and human rights: bridging the gap between responsibility and accountability, *Journal of Human Rights*, **14**, 237–259.
14. Wettstein, F. (2016). From side show to main act: can business and human rights save corporate responsibility?, in *Business and Human Rights: From Principles to Practice*, eds. Baumann-Pauly, D. and Nolan, J. (Routledge, London), pp. 77–87.
15. Wheeler, S. (2015). Global production, CSR and human rights: the courts of public opinion and the social licence to operate, *International Journal of Human Rights*, **19**(6), 757–778.
16. Mamic, I. (2005). Managing global supply chains: the sports footwear, apparel and retail sectors, *Journal of Business Ethics*, **59**, 81–100.
17. Hugill, A., Short, J. and Toffel, M. (2016). Beyond symbolic responses to private politics: examining labor standards improvement in global supply chains, *Harvard Business School Working Paper 17-001*, p. 33.
18. Nossar, I., Johnstone, R. and Quinlan, M. (2004). Regulating supply chains to address the occupational health and safety problems associated with precarious employment: the case of home-based clothing workers in Australia, *Australian Journal of Labour Law*, **137**, 17.
19. Nossar, I., Johnstone, R. and Quinlan, M. (2015). Protective legal regulation for home-based workers in Australian textile, clothing and footwear supply chains, *Journal of Industrial Relations*, **57**(4), 585–603.

Part II

INPUT-OUTPUT ANALYSIS

Chapter 3

Calculating the Cost of Trade

Arne Geschke
Integrated Sustainability Analysis, School of Physics, University of Sydney, NSW 2006, Australia
arne.geschke@sydney.edu.au

3.1 Introduction

The first two chapters of the book have already highlighted the importance of trade in today's world and how the final consumer of a good or service bears a certain responsibility for working conditions and other social aspects along the supply chain of these products and services. These social implications of international trade are receiving increasing interest in the public media, a prominent example being the *Guardian's* stories and discussions on slavery in the seafood supply chain that ran throughout 2016.[a] However, it is often the working conditions in specific factories or countries that are under the spotlight. Additionally, it often requires a tragic event, such as the collapse of a badly constructed factory building or an increased suicide rate among employees before attention is drawn

[a]https://www.theguardian.com/sustainable-business/2016/jul/18/slavery-in-the-seafood-supply-chain-what-can-we-do-about-it-event (accessed 23 November 2016).

The Social Effects of Global Trade
Edited by Joy Murray, Arunima Malik, and Arne Geschke

ISBN 978-981-4774-55-0 (Hardcover), 978-020-3728-71-0 (eBook)
www.panstanford.com

to working, and sometimes also living, conditions of the people who are directly affected. In these cases an individual supply chain that links the final consumers in developed countries to the place of the tragedy is uncovered and publicly discussed. Efforts are often made to rectify the situation for the local population that is directly affected (see, for example, the case of the Thai fishing industry, also reported in the *Guardian*, 2016). What often remains in the dark is that most products in our highly globalised world rely on thousands, if not millions, of supply chains. Social injustice can occur at any given point along any of these supply chains. The same story applies to social justice endeavours that occur along supply chains or the social good that comes from trade. However, we are perhaps even less likely to hear of the benefits of global trade, or of wrongs rectified, which undoubtedly do not have the urgency nor perhaps the same media appeal as its tragic consequences.

To fully understand the social effects of trade, one must identify and quantify all supply chains and the associated social impacts along them. This chapter discusses the information and computational requirements necessary to obtain a full global analysis in terms of both identification and quantification of social impacts embedded in supply chains.

Undoubtedly, this is an enormous task. Given the sheer volume of goods traded every day, understanding the complex network of global supply chains requires the gathering and analysis of huge amounts of data. One approach to master these extremely large amounts of data is to use input-output (IO) analysis[b], which was already briefly mentioned in the opening chapter of this book and which was also used to obtain the results that are discussed in the remaining chapters.

To use IO for global analyses, global IO databases must be available. Compiling global IO databases is a highly complex, labour- and data-intensive and also expensive undertaking. Additionally, it often still requires a certain amount of programming and IO skills to use these databases for analyses. Due to these constraints, only a handful of global IO databases are available today, and in the past these were mainly used by IO researchers.

To open up the potential of IO analysis to other research fields, the process of obtaining and using a large-scale IO model had to

[b]For a complete overview of IO analysis, see, for example, Ref. [1]. For a beginners' guide, see Ref. [2].

become faster and simpler. This chapter provides an example of a sophisticated IO research platform that was designed with these goals in mind—the Australian Industrial Ecology Lab (IELab)—developed by a collaborative research team in Australia with a grant from the Commonwealth Government 2012–2016.

For readers unfamiliar with IO analysis we explain briefly what it is and provide some of the history of IO development, from analysis of a single economy or region through to multi-regional analysis extending to global supply chains. Extensions of IO to include the environmental and social effects of doing business are then discussed, along with the concept of footprinting. From here the chapter takes a behind-the-scenes look at construction of an IO framework and the particulars of constructing IELab's unique collaborative and flexible research platform[c].

This chapter covers the concepts of IO analysis and IELab in the context of social footprinting at a purely conceptual level on the one hand, while on the other hand also sketching out the mathematical challenges for readers who wish to learn more about the nuts and bolts of IELab. This chapter has been written so that it can also stand without those sections covering the mathematical details. Readers are, therefore, encouraged to pick those sections that are most aligned with their interests and levels of mathematical proficiency.

3.2 Input-Output Analysis as a Tool to Analyse Global Supply Chains

The basic idea of IO analysis is to use direct transactions between pairs of directly trading entities in an economy and then use mathematical tools to chain these individual direct transactions into full supply chain networks. A simple example would be to record how much iron ore was bought by the steel industry and how much steel was bought by the car industry. These two individual direct transactions can then be combined to find out the dependency that

[c]The platform allows the user to choose the desired sectoral and regional detail (i.e., what goods and services are being traded, and where in the world these transactions are happening). Once these decisions are made, all available data are automatically converted, adjusted and reconciled to fit the defined multi-regional input-output (MRIO) structure. The user/researcher now has a custom-made IO table designed to address her or his specific research questions.

the car industry indirectly has on the iron ore industry. Obtaining these direct transactions is relatively straightforward: The accounting books of each company would have information about direct trade partners and the trade volumes. But to use these raw data for global supply chain analysis, one has to conceptualise the data acquisition for a number of reasons.

The first reason is that if every company's activity were to be considered individually, there would be too much data for today's computers to process all this information. Therefore, within IO analysis, the economy is organised into so-called sectors. Individual sectors comprise a defined set of products or services. All economic activities that fall within the scope of a sector are summarised within the sector. For example, the car-manufacturing sector would comprise all car construction activities within the economy. This can, of course, mean that more than one company is represented in a single sector. Similarly, an individual company may be represented in more than one sector if the company follows more than one commercial activity. Following the definition of all sectors, the monetary trade between each pairing of sectors is recorded in the so-called *intermediate transaction matrix*. This is an important feature of IO analysis: the trade between every possible pairing of sectors is recorded. For example, if steel and car manufacturing are two sectors in an intermediate transaction matrix, then the sale of cars from the car industry to the steel industry as well as the amount of steel bought by the car industry from the steel manufacturers are both recorded in individual monetary figures within this matrix.

The intermediate transaction matrix only holds the trade values between individual economic sectors. There are, however, a handful of agents that participate in the economy that cannot be assigned to individual sectors: households, government, capital and exports/imports. The interactions between these four agents and the economic sectors are recorded in separate matrices. The so-called *final demand block* contains the transaction values for all activities where the four agents receive goods and services from the economic sectors (such as a family buying a car), and the *value-added block* holds all transaction values where money flows from the economic sectors to the agents (such as families receiving an income from the steel industry). The three parts—intermediate transaction matrix, final demand block and value-added block—form the IO system or the IO table. Figure 3.1 depicts the structure of a basic IO system.

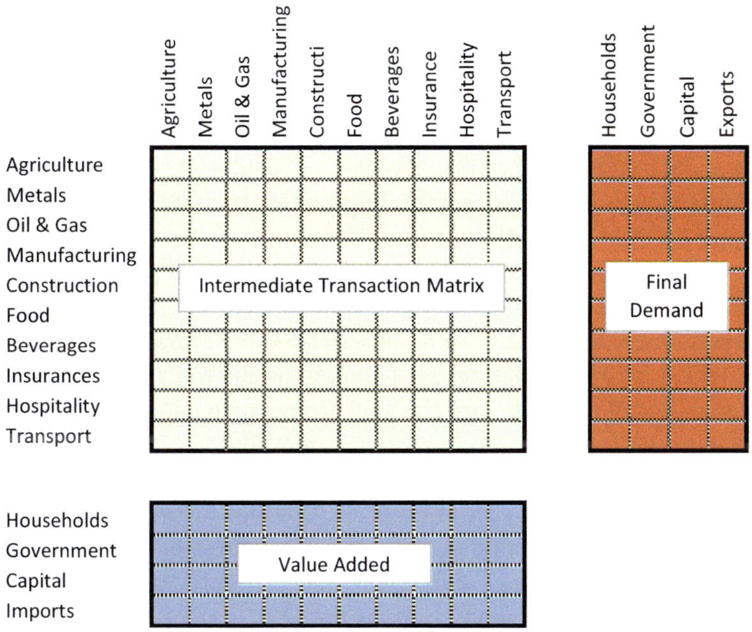

Figure 3.1 Basic IO system. The intermediate transactions matrix features 10 example sectors. The final demand and value-added blocks feature the four standard sectors for these blocks. The individual cells within the three matrices indicate where the transactions between the sectors are recorded.

The set of sectors used for a particular IO table is referred to as the IO classification. IO classifications often vary between different countries. Often, an IO classification shows more detail for those economic activities that are dominant in the economy that it is intended for. But other factors, such as available resources for compiling IO tables, also play a part when deciding on an IO classification.

So essentially, the basis for all IO analysis is large tables of numbers, and these numbers represent the direct transactions between different economic sectors or agents. The next step is to use these still fairly basic and intuitive pieces of information to obtain a full supply chain analysis.

IO analysis was first conceived by Wassily Leontief in the 1930s [3]. Leontief realised that by cleverly combining an IO system with specialised mathematical tools[d] [3], it is possible to extract information about the supply chains that are required to meet the final demand. Leontief saw that an IO table is relatively easy to construct. He further saw that the information about the complex network of supply chains that is required to produce a given final demand is hidden but available within an IO table. He developed the fundamental tools to extract this information from the IO tables. His work earned him the Nobel Prize in 1973.

While IO analysis was originally a tool developed solely to understand the economic interdependencies within an economy, it was soon realised that it could be extended in several ways. Those extensions that are important foundations for the subject of this book—social analysis—are the shift from single-nation (or single-region) IO tables to multi-regional input-output (MRIO) tables [4] and the inclusion of environmental data into IO analysis. In short, MRIOs develop the concept of IO tables further so that not only different economic sectors but also different geographical regions are considered. Extending this concept even further by considering environmental data in the MRIO table yields what is often called environmentally extended multi-regional input-output (EE-MRIO) analysis [5]. In EE-MRIO analysis, information about the economic structure of the supply chains is extracted from the MRIO framework. In a second step, this information is paired with data on the environmental pressures along the supply chain. This allows us to account for environmental impacts that occur along a network of supply chains that lead us inexorably to the final consumer (i.e., rather than the industries that directly cause these impacts). For example, the water use and greenhouse gas (GHG) emissions caused within a car factory are not accounted to the company running the factory but to the final consumers who purchase the cars. This technique is referred to as *footprinting*. There have been several landmark footprinting studies. For example, Hertwich and Peters [6] studied the carbon footprint of nations, Wiedmann et al. [4]

[d]The standard tools include footprint calculations, product layer decomposition and structural path analysis. A large number of variations of these methods are available to obtain and decipher information that is provided by IO tables.

worked on the material footprint of each nation and Lenzen et al. [7] uncovered the economic drivers for biodiversity threats.

To date, these environmental footprints have been the most popular footprint types. To calculate the social effects of trade, one must apply the footprinting technique not to environmental data but to social data instead. Social data have so far not received the same attention in MRIO analysis as environmental data have, and the research field of social footprinting is now developing rapidly.

To undertake a global social footprint analysis, the analyst requires a global MRIO table combined with social data for each region and economic sector represented in the MRIO. Constructing global MRIO frameworks at a high sectoral resolution has been a key topic of research in recent years, and several global MRIO frameworks are available today[e]. However, the sectoral resolution provided in each of these frameworks is fixed and cannot be changed by the user. Often, a particular research question calls for more or less sectoral detail for particular economic activities. Calculating a social footprint for garment-producing industries in the emerging economies of Southeast Asia, for example, would call for an MRIO framework that features a high sectoral detail for these industries, whereas the sectoral detail for the European service industries might not be too relevant. So why not simply construct a global MRIO framework that features high levels of regional and sectoral detail overall? Because it is close to impossible: the time and computing power required for such a task would be overwhelming because the construction of MRIO frameworks is computationally intensive. In this chapter we are, therefore, presenting a flexible approach to the construction of MRIO frameworks. The user can choose the desired sectoral and regional detail, and all available data are converted, adjusted and reconciled to fit the defined MRIO structure.

Offering such a high—and in the IO community unprecedented—level of variability is only possible if all raw data are seemingly integrated into a pre-defined data framework. The effort required to bring these raw data into the right shape can be significant. Additionally, building customised MRIO tables from these raw data

---

[e]See, for example, the WIOD database [8] (http://www.wiod.org/new_site/home.htm), the Exiobase project [1] (http://exiobase.eu) or the Eora database [9, 10] (http://worldmrio.com). See Ref. [10] for an overview of MRIO databases and case studies illustrating their use.

requires sophisticated mathematical methods combined with the use of high-performance computing architecture. The details of this process will be discussed in a separate section of this chapter.

3.3 Data Types and Bridging between Data of Varying Levels of Detail

We will now briefly focus on the challenges posed by the inclusion of social data into footprinting calculations, as well as the challenge posed by requiring IELab to allow both highly aggregated and highly detailed data to be considered.

Unlike social data, environmental data are usually expressed in physical units (such as tons of CO_2, megalitres of water and hectares of land). This property qualifies almost all environmental data for immediate use in IO analysis. Social data are much more diverse, and this poses challenges in terms of how to include these data into an IO framework. The property that all data within an IO framework must fulfil is that of additivity: If two sectors are merged into a single sector, all data that belonged to the individual sectors before must be added up in the new, merged sector. For example, assume the black-coal-mining sector employed 10,000 people and the brown-coal-mining sector employed 12,000 people. If one were to aggregate these two sectors into a single coal-mining sector, then the number of employees of this new sector would simply be the sum of the employees of the two previous sectors, namely 22,000.

But social data may also be presented in units that are not additive. An example for social data that is not by default fit for IO-related use is the percentage of unionisation of employees in a specific sector. These data do not fulfil the additivity condition. If 60% of black-coal miners and 70% of brown-coal miners are organised in unions one cannot automatically determine what the unionisation of all coal miners is. In this case the data must be translated into a unit that does allow for the adding up of individual sectoral data. For the unionisation example the percentage could be translated into the total number of employees who are organised in unions in each sector.

More complex social indicators can be accommodated in the framework but may require some sort of trade-off in order to fulfil the additivity condition. For example, in the case of estimating corruption by sector and country, national corruption indices can be converted into an additive quantity by multiplying with the employment data by sector and country. Corruption indices are provided in 'percentage of workers affected by corruption' [2]. Only one value is reported per country. To translate these measures into absolute numbers of corruption-affected jobs in each sector, we apply the country's percentage value to the number of total workers in each of its sectors. Thus, corruption is treated for each sector in each country as the number of jobs affected by corruption in that particular sector. Put simply, the corruption level in a given sector is defined by the number of workers in that sector and the country's overall corruption risk. More workers in a sector means more employment affected by corruption in that sector [2]. Of course this is an imperfect measure. However, researchers have always found ways to compensate for or accommodate less than ideal data and in explaining the assumptions behind their calculations they provide pointers to current issues, possible solutions and future research. In this way, researchers continually push the agenda forwards. The whole footprinting agenda, for example, has helped to drive changes and improvements in data identification and collection (albeit more so for the physical data of environmental indicators).[f] In the case of corruption footprints, an adverse result for a particular country may provide an incentive to push for data collection by sector, which would yield a more accurate result on which to base action. Whatever the indicator, when better data are available they can be used to build on and improve what has gone before.

The second big challenge for including social data into an IO framework is that different social data can call for vastly different regional and sectoral details in the IO framework. The International Labour Organization (ILO) publishes information on the number

---

[f]An example is the Global Footprint Network's National Footprint Accounts, which are updated annually with advice from the National Accounts Review Committee. See http://www.footprintnetwork.org/pt/index.php/GFN/page/footprint_data_and_results/ (accessed 24 November 2016).

of employees per economic sector for 22 sectors for each country,[g] which represents a highly aggregated classification. The other extreme is given by local surveys that provide very detailed information on labour issues within a small geographical boundary (like a factory, a town or a country) or a detailed survey for a multinational company that may have production plants in multiple regions of a specific country. The last two examples would require a very detailed regional and sectoral resolution in the IO framework. Due to the additivity of IO-compliant data it is, of course, possible to translate any source data to a more aggregated IO classification, but this measure usually implies the loss of detailed information. For the case of the local survey confined to a few towns, an IO framework that only allows for a regional resolution at the country level would render the outcomes of the survey most likely unusable because specific regional effects might be dwarfed by the potentially much larger masses of data relating to other regions of the same country. Due to the country-level representation, all of these source data would be represented in the same section of the MRIO and specialised regional effects captured by the local survey would be lost.

IELab was designed with this issue in mind. Given that IELab allows for MRIO tables that show high levels of detail for specific sectors or regions, it is important that the raw data repository and the mathematical methods that are used to generate the MRIO tables not compromise the much-sought-after details that regional studies can provide. At the same time IELab should also be able to reliably reflect high-level international trade relations, which form the backbone of any global trade analysis. IELab has to be able to 'marry' big data and small data into one framework. To achieve this, an artificial, highly detailed classification is used in which all raw data are saved. This root classification serves as a common meeting point for all data sources and ensures that the level of detail of any data source is maintained and no information is lost. The term 'root classification' was coined during the construction of the Australian Industrial Ecology Virtual Laboratory (AusIELab), during which the

[g]http://laborsta.ilo.org/applv8/data/isic4e.html (ISIC-Rev.4) (accessed 24 November 2016).

underlying concepts of variable, task-dependent MRIO frameworks were developed [11]. In essence, the root classification represents the highest level of detail for which data are available. For AusIELab, the root classification is derived from the 2009 household survey data published by the Australian Bureau of Statistics. Within this root classification, Australia is broken up into 2214 individual regions, with 1284 sectors for each region.

Most data sources offer information that is available at a lower level of detail than the root classification requires; therefore loss of detail is prevented.

3.4 Bringing It All Together: The Social Footprint Laboratory

As discussed before, in 2015–2016 AusIELab was the only operational IELab. However, the software engine that sits behind AusIELab was developed with future IELab-type applications in mind. Hence, the year following the launch of AusIELab several other IELabs were already under construction. These included labs for China, the Philippines and Indonesia. Ultimately this led to the creation of GlobalIELab, the aim of which was to feature a highly detailed root classification for every single country. GlobalIELab was designed to be capable of constructing highly detailed, customised global MRIO tables.

By January 2017, GlobalIELab was in an advanced developmental stage. As well as serving economic and environmental analysis needs it was designed to serve as the centrepiece for a global social footprint laboratory (SFL). The IELab software suite is highly modular and allows for the addition of new data at any time, which means that available social data can be integrated into GlobalIELab at any time. This feature extends the capability of GlobalIELab to also become a SFL. As discussed before, GlobalIELab can consider high-level global social data as well as regional, survey-based local data (provided the suitability of these data for IO-type analyses is ensured).

3.5 The Mathematical Task of Building a Model That Suits

This section describes the mathematical concepts that are required to enable the functionalities in IELab that were described and discussed before. This section is intended for the interested reader with a solid mathematical background and—to a certain extent—assumes that the reader is familiar with the underlying mathematical methods and terminology.

Before a purpose-built MRIO can be assembled, all data that are used for generating the model must be made available in the root classification. This does not necessarily mean that each data point is converted into this classification, but every data point must be understood if one is looking at it from the root classification. There are two ways data can be made available in the root classification:

- *Data conversion approach*: The data point can be explicitly translated into the root classification. This means that in many cases, the data points must be split into several smaller values and distributed across the corresponding sectors in the root classification. Splitting data points always requires additional information. For example, assume the data source reports a total value of 40 billion dollars for coal extraction but the root classification distinguishes between black-coal extraction and brown-coal extraction. To represent the source data in the root classification, the 40 billion dollars must be split and distributed across the two corresponding sectors black-coal extraction and brown-coal extraction. If no further information on coal extraction is available, this split must be based on assumptions, which waters down the reliability of the data source.
- *Metadata approach*: The second option is that rather than splitting each data point across all corresponding root classification sectors, meta-information about the data point is saved but the data point itself remains untouched. In the case of the coal extraction mentioned before, the system would save two pieces of information. The first piece would be the value 40 billion dollars and the second piece would be the information that this number represents the sum of

brown-coal and black-coal extraction. While this way of saving the data does not offer any information about the revenue of black coal, there are also no further assumptions made, that is, the quality of this information is not diluted in any way.

AusIELab uses both these data storage approaches. To construct the final MRIO, the raw data must firstly be translated from the root classification to the user-defined MRIO classification. Secondly the MRIO must be constructed from the prepared data. The key steps of the construction procedure are as follows:

1. Define the regional and sectoral classification for the MRIO.
2. Using data stored according to the data conversion approach, convert the data into the regional and sectoral detail chosen for the MRIO and assemble the initial intermediate transaction matrix $\mathbf{T}$, the final demand block Y and the value-added block V, and combine them into a single IO table.
3. Vectorise this IO table and denote the vectorised table $\mathbf{p}^0$.
4. Carry out inner consistency checks for the table by formulating and solving a number of linear equations that operate on the table elements. Use the metadata attached to the data stored according to the metadata approach to construct these linear equations in a coefficients matrix $\mathbf{G}$ and a right-hand-side vector $\mathbf{c}$ so that they match the sectoral and regional details chosen for the MRIO.
5. Find the final IO table $\mathbf{p}$ by solving the reconciliation problem $\min_{\mathbf{p}} f(\mathbf{p},\mathbf{p}^0)$ subject to $\mathbf{G}\mathbf{p} = \mathbf{c}$.

The final IO table is then obtained by reverting the vectorisation of $\mathbf{p}$. Geschke et al. [12] describe the mathematical reconciliation procedure in detail, and Lenzen et al. [11] outline the mathematical details to translate all data from the root classification into the MRIO classification as it is used in the IELab software suite.

Once the user has defined the regional and sectoral classification of the final MRIO, the remaining steps to construct the table, as outlined before, can be fully automated. Hence, once the source data are stored in the root classification, IELab makes the construction of purpose-built MRIO tables become a fully automated process.

A flowchart of the main steps to construct an MRIO within IELab is given in Fig. 3.2.

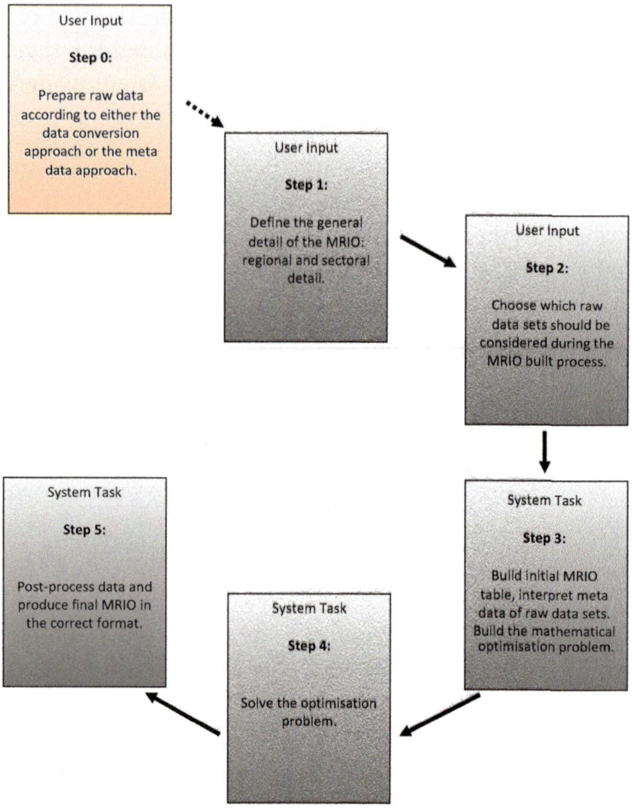

Figure 3.2 Flowchart of the IELab MRIO construction process. Steps 1–5 depict the standard MRIO construction process as it is carried out with IELab. Step 0 is the data preparation process to include any given data into the IELab system. This data preparation process does have to be carried out for each MRIO but is done prior to any MRIOs being constructed.

3.6 Conclusion and Outlook

This chapter describes the different concepts and data processing hurdles that must be mastered to calculate the social effects of trade. The concept of IO-based footprint calculations can easily be applied to social data once the data are in the right shape. The necessary tools to harness the information presented by global MRIO frameworks and social data have already been developed and tested within AusIELab. Additionally, GlobalIELab, a framework to port

AusIELab's capability to a global level, is being pieced together. By including social data into GlobalIELab, the calculation, analysis and assessment of global social impacts are finally possible. Calculation of the social effects of trade can now make use of the lessons learned from decades of EE-MRIO analysis and immediately join the group of matured methods used to analyse and understand the global impacts of our highly globalised economy.

References

1. Wood, R., et al. (2014). Global sustainability accounting: developing EXIOBASE for multi-regional footprint analysis, *Sustainability*, **7**(1), 138–163.
2. Xiao, Y., et al. (2017). The corruption footprints of nations, *Journal of Industrial Ecology*, doi:10.1111/jiec.12537.
3. Miller, R. E. and Blair, P. D. (2009). Foundations of input-output analysis, in *Input-Output Analysis: Foundations and Extensions*, ed. Blair, P. D. (Cambridge University Press, Cambridge; New York), pp. 10–68.
4. Wiedmann, T. O., et al. (2015). The material footprint of nations, *Proceedings of the National Academy of Sciences*, **112**(20), 6271–6276.
5. Tukker, A., Poliakov, E., Heijungs, R., Hawkins,T., Neuwahl,F., Rueda-Cantuche, J. M., Giljum, S., Moll, S., Oosterhaven, J. and Bouwmeester, M. (2009). Towards a global multi-regional environmentally extended input-output database, *Ecological Economics*, **68**, 1928–1937.
6. Hertwich, E. G. and Peters, G. P. (2009). Carbon footprint of nations: a global, trade-linked analysis, *Environmental Science & Technology*, **43**(16), 6414–6420.
7. Lenzen, M., et al. (2012). International trade drives biodiversity threats in developing nations, *Nature*, **486**(7401), 109–112.
8. Dietzenbacher, E., et al. (2013). The construction of world input-output tables in the WIOD project, *Economic Systems Research*, **25**(1), 71–98.
9. Lenzen, M., et al. (2012). Mapping the structure of the world economy, *Environmental Science & Technology*, **46**(15), 8374–8381.
10. Lenzen, M., et al. (2013). Building Eora: a global multi-region input-output database at high country and sector resolution, *Economic Systems Research*, **25**(1), 20–49.
11. Lenzen, M., et al. (2014). Compiling and using input–output frameworks through collaborative virtual laboratories, *Science of the Total Environment*, **485–486**, 241–251.

12. Geschke, A., et al. (2011). AISHA: a tool for constructing time series and large environmental and social accounting matrices using constrained optimisation, *19th International Input-Output Conference*, Alexandria, USA, http://www.iioa.org/Conference/19th-downable paper.htm.

Chapter 4

Human Rights Due Diligence with the Social Hotspots Database

Catherine Benoît Norris, Gregory Norris, Yanyan Xiao, and Joy Murray
New Earth, Harvard Extension School, Harvard School of Public Health, University of Sydney, NSW, Australia
catherinebenoit@fas.harvard.edu; gnorris@hsph.harvard.edu

4.1 Introduction

When the United Nations (UN) *Guiding Principles on Business and Human Rights* were endorsed by the UN Human Rights Council in 2011, former special representative of UN Secretary General Professor John Ruggie proclaimed that the era of declaratory corporate social responsibility (CSR) was over. What he meant was that voluntary CSR was no longer an option and that companies now had to develop and continuously implement a process of *due diligence*. This expression was coined after the term used in the financial sector. It means that all reasonable steps are taken to verify the good standing of an investment.

Human rights due diligence is thus a reasonable investigation of the human rights risks that may be present in a company's supply

The Social Effects of Global Trade
Edited by Joy Murray, Arunima Malik, and Arne Geschke

ISBN 978-981-4774-55-0 (Hardcover), 978-020-3728-71-0 (eBook)
www.panstanford.com

chains. The objective of the guiding principles was to develop a framework that would become part of international law and eventually be embedded in countries' legislations. Since then, a growing number of countries have passed laws that integrate the guiding principles' recommendation of requiring human rights due diligence. Within Europe, France vigilance law, the UK Modern Slavery Act, the Swiss referendum for due diligence, the Netherlands Child Labour Due Diligence Law and the EU non-financial reporting directive all seek to foster human rights accountability within global supply chains. In North America, the US has the Trade Facilitation and Trade Enforcement Act (H.R. 644), with Section 910 having been fortified to empower restrictions on the import of goods produced with forced labour.

To renew global collaboration to improve human well-being worldwide and following global consultation, the UN launched the Sustainable Development Goals (SDGs) in 2015.

The SDGs provide a compelling vision of humanity's progress that can inspire citizen actions as well as policymaking and business within and between countries. This long-awaited road map connects with important instruments, such as the Paris climate agreement and the UN guiding principles, and breathes new life into the UN's *Universal Declaration of Human Rights*. As the Commission for Business and Sustainable Development argues, 'The Sustainable Development Goals (SDGs) present an opportunity not just to update our vision of the role of business in sustainable development, but to change it fundamentally. There is no more pressing or more powerful way for business to accelerate social development than by driving respect for human rights across their value chains' [1].

Therefore, the wave of legislations related to the guiding principles, coupled with strong corporate support for the SDGs, is now compelling companies to establish a process to learn about, prioritise and act upon their supply chain risks. One method perfectly suited for this type of application is still under the radar. It is known as the *social life cycle assessment* (SLCA), one of the tools in the life cycle management toolbox. The SLCA is a technique to assess the social impacts of products and organisations from extraction of raw material to end of life. It has been developed from environmental life cycle assessment, the method used to conduct carbon and water footprints, for instance. What the SLCA can offer is a systematic assessment framework. To operationalise the SLCA an assessment system composed of methods, models and data, called the Social Hotspots Database (SHDB), was created in 2009 and made available publicly starting in 2013.

The preceding chapter has offered a detailed account of the information and computational requirements necessary to obtain a full global analysis of trade. We will now discuss how governments and organisations can use global input-output (IO) models along with social impact information to carry out due diligence and calculate a social footprint (SF) in the context of the SDGs and the UN Guiding Principles. To illustrate the process, we will use the example of gender equality. But first let's describe the SHDB assessment system.

4.2 Structure of the SHDB

One of the big challenges for organisations conducting human rights due diligence is the lack of a systemic and systematic assessment framework enabling quantification. This is needed to prioritise higher-risk issues, countries and supply chain activities.

Qualitative supply chain models are, hence, often being used. These are built from the ground up, by way of interviews and surveys. A qualitative supply chain model and human rights assessment are required for each commodity or good. The results of such studies are very detailed but fall short of answering the question, Where should a company invest resources to address the more salient problems, overall? In addition, large organisations that sell, invest or produce in a diverse set of product lines and portfolios have little means to identify supply chain overlaps and prioritise which countries, supply chain activities and issues they should work on first.

To answer these questions, an organisation/government needs a top-down assessment system that allows it to calculate the total SF of its purchases and the contributions made by each sector/commodity to the total social risk in a systematic and quantitative way. This is what the SHDB is meant to deliver.

The SHDB is designed to be a modular system, which includes three main components:

- A trade or supply chain model such as a global IO model

- A worker-hour model
- Data on social risks and opportunities

Technically, the SHDB is an extended IO life cycle inventory database providing a solution to enable (i) the modelling of product systems and (ii) the initial assessment of potential social impacts. The SHDB contains data for 57 different sectors, in each of 227 countries and territories. Thus, there are 12,939 unit processes in the database [2].

Although the SHDB was initially developed using a multi-region input-output (MRIO) model based on the Global Trade Analysis Project (GTAP) Version 7, the global economic equilibrium model [3], it can now be used with a variety of trade and supply chain models, including Eora, the World Input-Output Database (WIOD), and process-based models, such as Ecoinvent. In particular, a collaboration between New Earth and the University of Sydney has resulted in the development of an SHDB Eora version.

The Eora MRIO model is similar to GTAP, however, in more regional and sectoral detail, covering monetary and physical data for 189 countries at a detail level of between 25 and 500 sectors by country. The version of the SHDB that works in combination with Eora maps [7] social risk data to Eora higher sectoral resolution when the additional granularity is available.

The second piece of the assessment system is a worker-hour model. Labour hour intensity factors are used together with the social risk–level characterisations to express social risks and opportunities in terms of work hours, by sector and country, at a given level of risk relative to each of over 22 social impact sub-categories and over 150 different indicators. The risk data address five main impact categories: labour rights and decent work, human rights, health and safety, governance and community in line with major social responsibility references [4].

The SHDB project draws upon hundreds of data sources, including the International Labour Organization, the World Health Organization, the US Department of Labor and State and the World Bank.

Quantitative statistics and qualitative information by country and sector are used to develop characterisation models. These models assign a risk (or opportunity) level to the data (low, medium, high, very high, no data and no evidence) so that users can identify target areas in their supply chains to verify or improve social conditions. This consistent assessment framework allows risk comparison to be made between countries and sectors.

The SHDB is based upon life cycle attribute assessment (LCAA), a methodology developed by Norris [5]. Each unit process (which we define as a country-specific sector when using global IO models) has a number of different attributes, or characteristics, relative to a large set of social issues. The activity variable used in the SHDB (and in the SLCA literature) is worker-hours. Thus, the SHDB can be used to identify how many worker-hours are involved for each unit process in the supply chain for a given final demand (final product or service output from the system). The sociosphere flows are expressed as worker-hours at a specified level of risk on a given risk indicator per US dollar of process output.

Worker-hours are relevant because they represent evidence of the intensity of work required by each country-specific sector directly related to production. Work intensity is one of the criteria proposed to prioritise decision and action. Furthermore, if work intensity is important in a specific country and sector, not only the impacts of the stakeholder category for workers may be important but also the impacts affecting all other categories of relevant stakeholders (local community, society, supply chain actors).

Even though worker-hours may be less directly linked to issues related to local communities and society, they remain to date the only activity variable that can be used to assess the scale of an issue within the context of the supply chain as a whole.

4.3 Integrating the Social Hotspots and Eora Databases

To create an SHDB expressed in the Eora MRIO regional and sectoral classification, we follow the approach noted above, where social risk levels are expressed in terms of worker-hours by sector and country (the *activity variable*). To do so, we use worker-hour data obtained from Eora's employment satellite account. In particular, we weight qualitative social risk data from the SHDB with Eora's employment data to build a social risk satellite account for Eora.

Five main steps are followed to obtain the Eora social risk account. The first step is to extract the world employment satellite account from the Eora MRIO database. The data in this account are measured in units of full-time equivalent (FTE) person-years by sector and country. For each country listed in Eora we also extract its risk data (by country and sector) from the SHDB. These risk data are expressed in qualitative format, including six levels: very high, high, medium, low, no data and no evidence.

The second step is to reclassify the extracted table to convert the GTAP-classed qualitative risk data in the SHDB into quantitative data that can be used to populate a satellite account fit for the Leontief calculus.

The third step is to convert the qualitative risk matrix into a binary matrix that has every one of the original 155 indicator rows split into six rows and was hence re-sized. The six new rows correspond to the risk levels very high, high, medium, low, no data and no evidence.

The fourth step is to convert the binary matrix into units of worker-hours, simply by element-wise multiplication of every row by the employment account created in the second step.

Finally, an Eora-classed matrix expressed in units of worker-hours is obtained once again according to the mapping procedure described using the concordance matrix *C* from the second step and the Eora-classed total output as a proxy variable for normalisation.[a]

4.4 Calculating a Social Risk Footprint

The SHDB risk tables characterise social risks within a country and within an economic sector. By allocating social risk to countries, the SHDB follows a territorial perspective, equivalent to carbon emissions inventories of countries submitted under the United Nations Framework Convention on Climate Change (UNFCCC). By allocating risk to producing sectors, the SHDB follows a producer perspective. In contrast, using the SHDB with a supply chain model and work intensity information (LCAA) allows one to calculate a social risk footprint that characterises entire supply chains involving many sectors and many regions. Its supply chain

[a]These five steps are adapted from Ref. [6].

scope is equivalent to that of the carbon footprint of products and countries [7].

We calculate an SF by subjecting the unweighted and risk-weighted satellite accounts to the Leontief calculus.[b] This provides us with two employment footprint representations for an arbitrary final demand bundle, the *unweighted employment footprint* and the *risk-weighted employment footprint*. The ratio of these two defines the social risk footprint of region-sector pairs for each indicator. The purpose of the weights is to allow additivity, necessary to calculate a footprint, while putting a greater emphasis on country-specific sectors and supply chains at higher risks of the social issues under evaluation (see Table 4.1).

Table 4.1 Weights applied to the risk-weighted employment footprint

Risk level	**Weight**
Very high	0.9
High	0.8
Medium	0.5
Low	0.2
No data	0.5
Not applicable/No evidence	0.5

By facilitation of the Leontief inverse, the social risk footprint includes both direct and indirect supply chain risks.

Our approach to describe a risk footprint as a ratio of the footprints of a weighted and an unweighted activity variable (here employment) gets around the problem that the risk measure as such is not additive (e.g., a low risk of 0.2 in two sectors and a medium risk of 0.5 in another sector do not add up to an overall very high risk of 0.9) and is, therefore, not suited to populate a satellite account for inclusion in a Leontief-type footprint calculus. This ratio method has been used previously where the non-additive Gini index was successfully incorporated into an inequality footprint by using separate income and employment footprints in a Lorenz curve calculation [8].

[b] $L = (I-A)^{-1}$ is called the Leontief inverse; in addition to the direct transactions in A, it captures the indirect supply chain links between producing and physically impacting sectors and distant consumers.

4.5 Applying the SHDB to Carry Out Human Rights Due Diligence

The guiding principles call on organisations to carry out human rights due diligence on their operations as well as their trade/business relationships. Previous studies have demonstrated that developed economies' SFs are often driven by their imports while developing economies' own activities contribute significantly to their SFs, sometimes in addition to their imports [8–10]. We consider an SF to be the measurement of the negative social impacts associated with the supply chain activities required to deliver a product or fulfil an organisation's mission or a country's needs. In this study, we illustrate how SF calculations can be a tool applied in the context of a due diligence program. We focus on gender equality to illustrate the power of a combined SHDB [11] and Eora database of the global economy to show the in-depth intelligence provided by this high-level combination of data sources.

Using the Eora MRIO and SHDB combination, we produce a quantitative, consistent and reliable measure of the SF of nations as an example of the calculation of the SF of any organisation. Selection of the indicators that make up our SF was influenced by the UN's final report on the Millennium Development Goals (MDGs) [12], which stated gender equality and empowerment of women must be at the heart of the SDGs. Gender equality is also one of the human rights issues considered in a human rights due diligence. Having singled out this specific area of concern as a priority for the SF our next step was to examine the indicators supported by the SHDB. We identified four SHDB indicators that either directly addressed or contributed to gender equality. These became the components of our SF.

The SHDB includes over 150 indicators. From these we selected four indicators that match the SDGs that we wanted to capture in our SF calculation (see Table 4.2). We use the gender equality indicator to show three detailed findings: (i) ranking of countries by net importers and net exporters of gender inequality risk, (ii) a map of global flows of gender inequality risk embodied in international trade and (iii) details of trading partners and commodities shown in the map.

Table 4.2 Selection of gender equity risk indicator and related SDG[c]

Contributing SHDB indicators	SDG
1. Gender equality	
83. Human Rights: Gender Equity: Risk of gender inequality by sector based on representation in the workforce	**Goal 5 Gender equality** 5.1 End all forms of discrimination against all women and girls everywhere.
77. Human Rights: Gender Equity: Characterization of the World Economic Forum's Global Gender Gap Index	**Goal 5 Gender equality** 5.c Adopt and strengthen sound policies and enforceable legislation for the promotion of gender equality and the empowerment of all women and girls at all levels.
23. Labor Rights & Decent Work: Migrant Workers: Risk that women are not accepted into the country as immigrants	**Goal 8 Decent work and economic growth** 8.8 Protect labor rights and promote safe and secure working environments for all workers, including migrant workers, in particular women migrants, and those in precarious employment.
78. Human Rights: Gender Equity: Characterization of the United Nations Development Program's Gender Inequality Index	**Goal 8 Decent work and economic growth** 8.5 By 2030, achieve full and productive employment and decent work for all women and men, including for young people and persons with disabilities, and equal pay for work of equal value.

4.6 Human Rights Due Diligence: The Case of Gender Inequality Risk

To calculate the SF of a country we need data or estimates of imports in a currency such as the dollar. Countries can be net importers or exporters of social risks. If the risks associated with their imports

[c]Adapted from Ref. [6].

outweigh the risks associated with their exports, countries are considered as net importers of social risk.

Figure 4.1 shows top- and bottom-ranking countries in terms of their footprint-to-domestic-social-risk ratio for the gender equality indicator. Bottom countries face more risks at home and can be seen as top exporters of gender equality risks, while top countries are net importers of social risk since they import goods that have a higher level of gender equality risk.

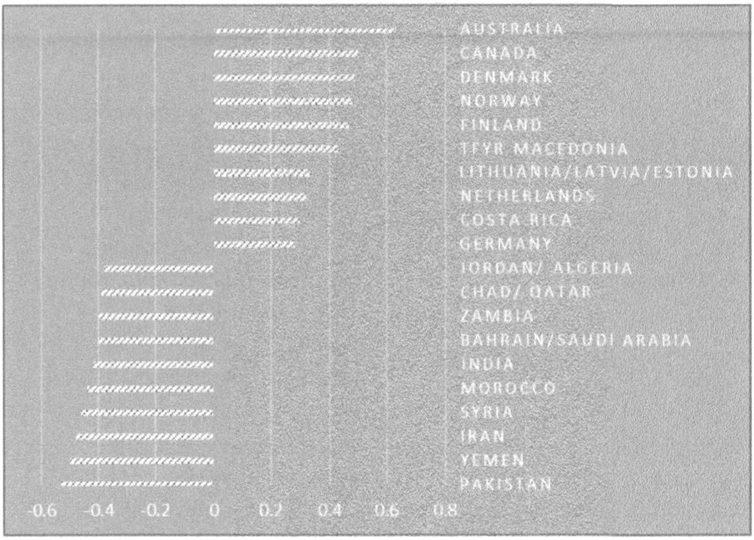

Figure 4.1 Net exporters and importers of gender inequality risk.

To calculate the SF of an organisation, we need data or estimates of purchases (in a currency such as US dollar) by input category (such as commodity or economic sector) and country. With this information, we can calculate the total size of the year's supply chain SF by multiplying the worker-hours at risks associated with the purchases by the risk-level weight (by country and economic sector) for each impact category and in total.

The calculated magnitude of social risk caused by international trade is determined by a combination of risk level and size of international shipments. This can be conveniently depicted as a world flow map of social risk embodiments in units of risk-weighted employment (Fig. 4.2). The same can be calculated for an organisation.

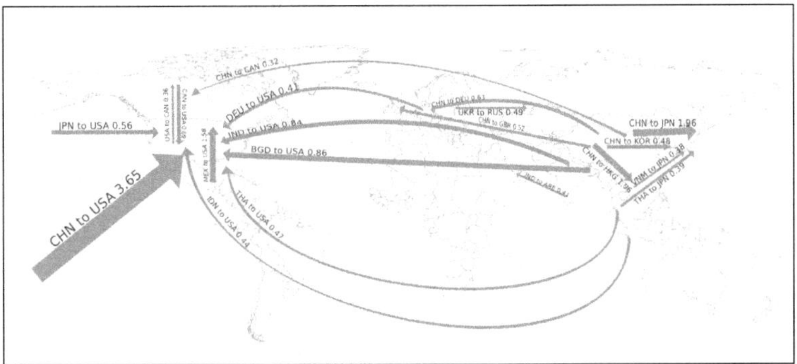

Figure 4.2 Social risk embodiments by trade flow. Reprinted from Ref. [6], Copyright (2017), with permission from Elsevier.

In this picture, the United States–Canada trade link conveys a sizeable social risk because the trade flow between these two neighbours is large. The same holds for the Japan–United States and Germany–United States links. These, however, form exceptions, and the major part of the flow pattern is dominated by social risk affecting people in China, Thailand, India, Mexico, Bangladesh and Indonesia, involved in producing export commodities, ultimately for the benefit of consumers in the United States, Germany or the United Kingdom.

We refer to the notion of embodiment to allude to consumers' implication in social issues arising in developing countries that produce their consumer items. We examined the overall SFs by indicator, country and bilateral trade relationships. The next figure (Fig. 4.3) presents the gender inequality footprint for the most important importers. The countries examined are largely identical with the top-ranking countries in Fig. 4.1. The results bear witness to the size of the total imports for each market and the risks related to where the imports are coming from and from which sectors. For instance, the high US gender inequality footprint reflects that it is the world's largest economy but also that it may import goods at higher risk of gender inequality.

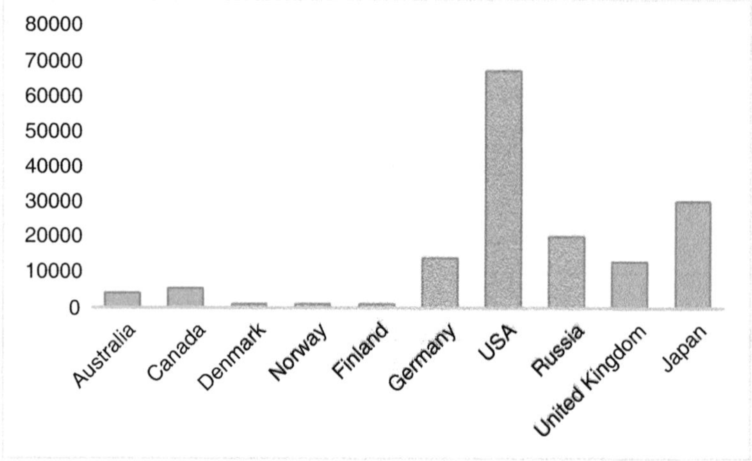

Figure 4.3 Total gender inequality footprint (full-time equivalent [FTE]).

Of all the countries shown above, China is the first or second main contributor to the gender inequality footprint, again because of the size of the exports to each of the countries but also because of the significant risk associated with this country's sectors, especially clothing. Sector wise, clothing, tourism, leather, crops (coffee and cocoa) and cars and motor vehicle parts are frequently topping the list of the main contributors to the gender inequality footprint.

Next, we show the country source and type of imports that contribute most to the gender inequality footprint for some of the highest-ranking countries: the United States (Fig. 4.4), Japan (Fig. 4.5) and Germany (Fig. 4.6).

The US gender inequality footprint stems from large quantities of imports from China, in particular the clothing and communication equipment sectors; Mexico, especially from cars, motor vehicle parts and clothing imports; Bangladesh, from the textile, transport and crustacean sectors; India, from the petroleum oil, clothing and medicament sectors; and Canada, from motor vehicle part, textile and car sectors. Although it is unlikely that the gender risk associated with Canada's imports will be very high (even if the sectors identified are likely to be less gender-equal than others), the large quantities of imports from Canada to the United States is why Canada's imports are included among the main contributors to the footprint.

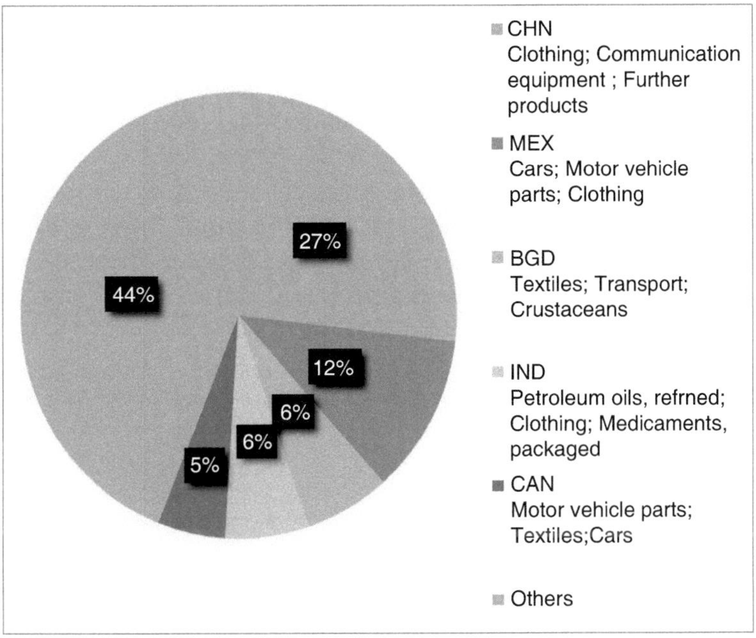

Figure 4.4 US imports' contribution to the gender inequality footprint.

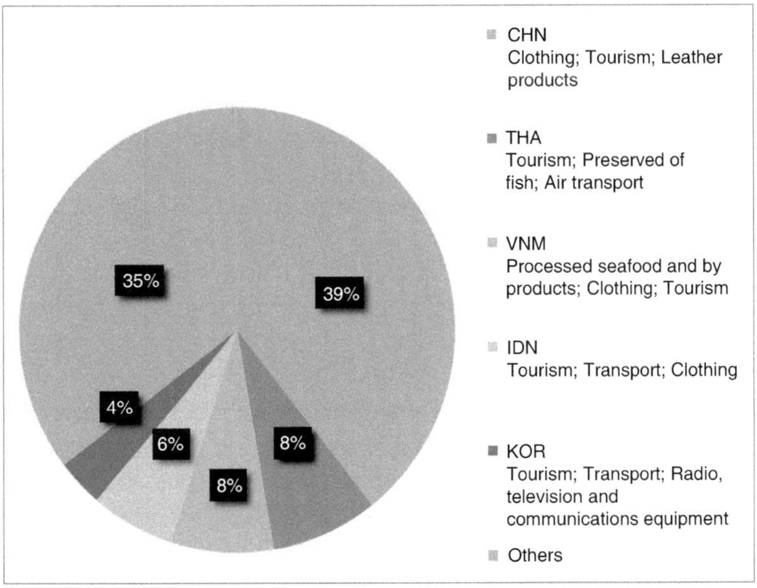

Figure 4.5 Japan imports' contribution to the gender inequality footprint.

Japan's gender inequality footprint is caused by imports from China, especially clothing, tourism and leather products; Thailand's tourism, preservation of fish and air transport; India's tourism, transport and clothing sectors; and South Korea's tourism, transport and communication equipment sectors (Fig. 4.5).

While the United States and Japan had China's sectors contributing a large percentage to the gender inequality footprint, Germany has a smaller share of its imports risk contributed by China, which shows a major difference in its import pattern (Fig. 4.6).

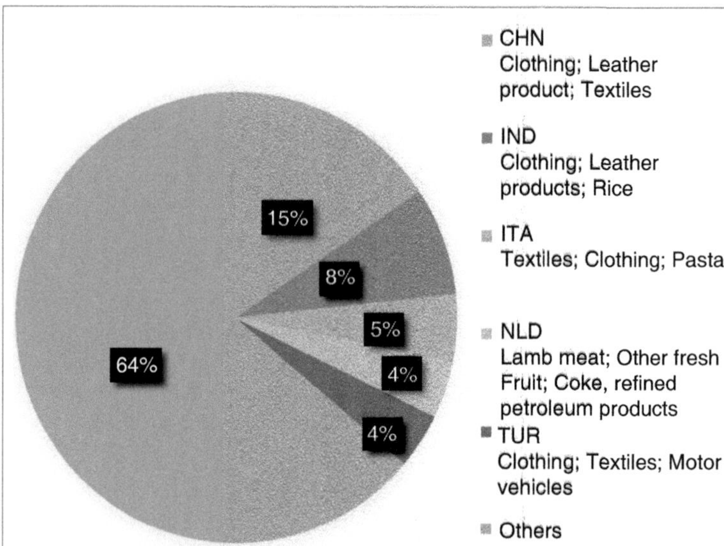

Figure 4.6 Germany imports' contribution to the gender inequality footprint.

These assessments demonstrate that (i) there is overlap in the main countries and sectors driving the gender inequality footprint (e.g., China: clothing and textiles) and also that (ii) each country may have, in addition, a mix of different countries and sectors that also contribute a significant share to the gender inequality risk (e.g., Germany has a mix of European countries contributing to its risk). Therefore, there is an argument both in favour of concerted efforts and in favour of country-specific processes to address gender inequality risks.

Our work highlights another issue. Figures 4.2 and 4.3 show significant flows wherever either the trade flow is high or the gender risk is high. In the first case the gender risk can be minor but magnified by the volume of trade. In the second case the volume of trade can be low but magnified by the seriousness of the risk. One way to put greater value on the high risk of gender inequality is to apply weight that captures more dramatically the potentiality for an adverse situation to occur. For example, weights in the SHDB default impact assessment method multiply by 10 the very high risks, by 5 the high risks, by 1 the medium risks and by 0.1 the low risks. This would lessen the magnification of the gender risk related to the volume of trade.

Even though this type of assessment sheds light on where high-risk issues are located and in which economic sectors, it does not point to the root causes. For instance, countries that are net exporters of social risks may be locked into risky export activities in order to meet their debt repayment obligations or because transnational corporations, seeking an ever-cheaper workforce, have no interest in improving conditions [13]. Such situations require appropriate trade policies; however, even though environmentally risky exports such as raw materials from extractive industries have long been recognised, addressing such issues from within the exporting country is not easy. The power differential between those who profit from the industry and those who work in it is huge [13]. In the case of social risk, sometimes it has required action on the consumer/importer side, for example, Section 1502 of the US Dodd–Frank Act for reporting on use of conflict minerals or the new EU Conflict Mineral Regulation, to bring about changes in working conditions. Future supply-side measures may be made possible by debt relief programs and debt management as those currently being undertaken by the World Bank Group [14]. These measures will address infrastructure gaps and invest in human capital [15] to assist in capacity building for environmental and socioeconomic policy development.

4.7 Conclusion

Governments and organisations can use global IO models along with social impact information to carry out human rights due diligence on their purchase and import supply chains and calculate an SF. Doing so can help them focus resources on the most salient production activities and countries where the largest share of their risk lies.

The UN guiding principles call for companies to implement a human rights due diligence process that encompasses all their business relationships and supply chains. To effectively understand the risks associated with the whole of their supply chains, organisations require a top-down model. Specific information on suppliers' location and performance can then be sought in an organised fashion, prioritising higher-risk business relationships and supply chain links.

This way, tracking improvements in the context of a human rights due diligence program or so that stakeholders everywhere can know with confidence that the SDGs are being met can become possible.

Two additional questions may then arise: What are the root causes of these adverse social impacts, and what are the most promising improvement opportunities?

References

1. BSDC and Shift. (2016). *Business, Human Rights and the Sustainable Development Goals Forging a Coherent Vision and Strategy* (BSDC and Shift, London).
2. Benoît Norris, C., Cavan, D. A. and Norris, G. (2012). Identifying social impacts in product supply chains: overview and application of the Social Hotspot Database, *Sustainability*, **4**(9), 1946–1965.
3. Center for Global Trade Analysis Department of Agricultural Economics Purdue University. (2008). *Global Trade Analysis Project (GTAP)*, https://www.gtap.agecon.purdue.edu/.
4. Lehmann, A., Zschieschang, E., Traverso, M., Finkbeiner, M. and Schebek, L. (2013). Social aspects for sustainability assessment of technologies: challenges for social life cycle assessment (SLCA), *International Journal of Life Cycle Assessment*, **18**(8), 1581–1592.
5. Norris, G. A. (2006). Social impacts in product life cycles towards life cycle attribute assessment, *International Journal of Life Cycle Assessment*, **11**(1), 97–104.
6. Xiao, Y., Benoît Norris, C., Lenzen, M., Norris, G. and Murray, J. (2017). How social footprints of nations can assist in achieving the Sustainable Development Goals, *Ecological Economics*, **135**, 55–65.

7. Wiedmann, T. (2009). Carbon footprint and input-output analysis: an introduction, *Economic Systems Research*, **21**, 175–186.
8. Alsamawi, A., Murray, J. and Lenzen, M. (2014). The employment footprints of nations, *Journal of Industrial Ecology*, **18**, 59–70.
9. Lenzen, M., Kanemoto, K., Moran, D. and Geschke, A. (2012). Mapping the structure of the world economy, *Environmental Science & Technology*, **46**(15), 8374–8381.
10. Simas, M. S., Golsteijn, L., Huijbregts, M. A. J., Wood, R. and Hertwich, E. G. (2014). The 'bad labor' footprint: quantifying the social impacts of globalization, *Sustainability*, **6**(11), 7514–7540.
11. Benoît Norris, C. and Norris, G. (2015). The social hotspots database, in *The Sustainability Practitioner's Guide to Social Analysis and Assessment*, eds. Murray, J., McBain, D. and Wiedmann, T. (Common Ground, USA).
12. United Nations. (2015). Universal Declaration of Human Rights, United Nations.
13. Muradian, R. and Martinez-Alier, J. (2001). Trade and the environment: from a 'Southern' perspective, *Ecological Economics*, **36**, 281–297.
14. Battaile, B., Hernandez, R. and Norambuena, V. (2015). *Debt Sustainability in Sub-Saharan Africa Unraveling Country-Specific Risks* (World Bank Group, Macroeconomics and Fiscal Management Global Practice Group).
15. International Development Association and International Monetary Fund. (2011). *Joint IMF/World Bank Debt Sustainability Analysis 2011, IDA & IMF.*

Part III

Case Studies

Chapter 5

Corruption Embodied in International Trade

Yanyan Xiao
ISA, School of Physics A28, University of Sydney, Sydney, NSW 2006, Australia
yanyan.xiao@sydney.edu.au

5.1 Introduction

The commonly accepted definition of corruption is 'the abuse of public power for private benefit'. Corruption exists in all countries and at all levels of society and government. It is one of the biggest global problems, which affects a number of economic behaviours. International trade, as a primary accelerator of economic growth, is generally hampered by corruption [1]. Some researchers have studied the relationships between international trade and corruption within countries [2–5]. Furthermore, Knorich [6] finds that the corruption levels in different countries are correlated, indicating that corruption occurring in one country may have repercussions for corruption in other countries. These cross-country interactions between trade and corruption suggest that domestic corruption

The Social Effects of Global Trade
Edited by Joy Murray, Arunima Malik, and Arne Geschke

ISBN 978-981-4774-55-0 (Hardcover), 978-020-3728-71-0 (eBook)
www.panstanford.com

levels can be influenced through trading with more or less corrupt countries.

Adopting multi-regional input-output (MRIO) analysis that has been employed for analysing the supply chain effects of international trade on various environmental and social indicators, we calculate corruption footprints of 189 countries and regions and show commodity details for corrupt sectors. This comprehensive picture of global corruption shows the linkage between corruption and international trade from a customer responsibility aspect. Our results provide customers with full knowledge of the responsibility they share in indirectly supporting corruption when purchasing certain goods. Therefore, individually, customers can use these results as a guide to make responsible consumption decisions to ameliorate corruption elsewhere, for example, through using purchasing power to buy products only from corruption-clean international companies. Collectively, customers, community groups, and non-governmental organisations (NGOs) could pressure global brands and governments to develop anti-corruption policies to target industries wherever corruption is found in the supply chain.

5.2 Methodology

To calculate the corruption footprints of nations, the MRIO framework needs to be extended with a corruption satellite account. The most widely accepted approach to corruption measurement is that of *perceptions* of corruption. By definition, corruption perception indices are subjective and thus cannot depict accurate corruption levels in every economic sector across a country's entire economy.

Following the rationale used by the creators of the Social Hotspot Database at Harvard University [7, 8], we use corruption perception indices κ as national corruption weights to scale the employment data E. Then the estimation of corruption-weighted employment K ($K = E \times \kappa$) can act as a measurement for the corruption levels by sector and country, which is, namely, the corruption satellite account. Then K is subjected to the demand-pull calculus of an extended input-output model for corruption footprint calculations. In this work, we have calculated corruption footprints of 189 countries and regions in 15,979 industry sectors. Furthermore, we provide the commodity details of corrupt sectors in top importers of corruption.

The three most widely used corruption perception indices, the Corruption Perception Index (CPI), the Control of Corruption Index (CCI) and the Irregular Payments and Bribe (IPB) Index, are extracted from Transparency International [9], the World Bank [10] and the World Economic Forum [11], respectively, and converted in the same scale from 0 to 100 as in the current CPI (where 0 equals the highest level of perceived corruption and 100 equals the lowest level of perceived corruption). The employment data measured in full-time equivalent (FTE) person-years by sector and country are gained from the International Labour Organization's Labour Statistics (LABORSTA) database and the United Nations System of National Accounts Official Country database. The interindustry transaction data are downloaded from the MRIO database Eora (www.worldmrio.com).

5.3 Results and Discussion

5.3.1 Comparison of Corruption Situation Measured by Corruption Perception and Corruption Footprint

The corruption perception indices show the corruption situation within countries; they treat every country as an enclosed environment for corruption. However, corruption footprints show the corruption situation by taking international trade influences into consideration; they treat the entire world as an integrated environment for corruption. We provide a comparison of the corruption situation measured by the CPI and the corruption footprint in order to disclose how international trade affects the corruption status among countries.

Every country's corruption footprint is defined from a consumption-based perspective, which includes its domestic corruption and the imported corruption from all other countries embodied in goods and services required for meeting its final demand. The net corruption trade in one country is defined as its total imported corruption minus the total exported corruption.

The top and bottom 10 corrupt countries and regions measured in the CPI are shown in Table 5.1. The top net corruption importers and top net exporters are shown in Table 5.2. Comparing these two tables, we can see from the consumption-based view, the corruption status among countries has changed. This is because a country's imported corruption is affected by the corruption levels of its trading countries. For example, Hong Kong tops the list of net corruption importers, and the net imported corruption amount casts all the other importers far behind. It implies two messages: One is the domestic corruption in Hong Kong is very low, which means Hong Kong is very clean as a separate region; the other one is the corruption levels in its trading countries are much higher than Hong Kong's.

Table 5.1 List of top and bottom 10 corrupt countries and regions*

		Corruption Perception Index
1	New Zealand	95
2	Finland	94
3	Denmark	94
4	Sweden	93
5	Singapore	92
6	Norway	90
7	Netherland	89
8	Australia	88
9	Switzerland	88
10	Canada	87
11	Iraq	18
12	Haiti	18
13	South Sudan	16.4
14	Uzbekistan	16
15	Turkmenistan	16
16	Sudan	16
17	Myanmar	15
18	Afghanistan	15
19	Somalia	10
20	North Korea	10

*The lower the mark in the CPI, the more corruption in the country (e.g., North Korea is perceived as the most corrupt country in the world).

Table 5.2 List of top 10 net corruption importers and bottom 10 exporters*

		Net corruption imports/1000 capita
1	Hong Kong	1.0065
2	Kuwait	0.3078
3	United Arab Emirates	0.1984
4	Switzerland	0.1357
5	Australia	0.1170
6	Norway	0.1054
7	Singapore	0.0961
8	Qatar	0.0758
9	America	0.0720
10	United Kingdom	0.0713
11	Vietnam	−0.0198
12	Guatemala	−0.0258
13	Malaysia	−0.0260
14	China	−0.0269
15	Uganda	−0.0275
16	Thailand	−0.0369
17	Tanzania	−0.0422
18	Cambodia	−0.0565
19	Tajikistan	−0.0645
20	Madagascar	−0.0650

*The top 10 countries are net importers, while the bottom 10 countries are net exporters of corruption. This table is an excerpt (the last column) from Table S4 in Ref. [12].

International trade is an economic pillar in Hong Kong and earns significant benefits for this special administrative region of the People's Republic of China. However, international trade also brings obvious social risks, making Hong Kong, with its large amount of imported corruption, a vivid example of the social costs embodied in international trade.

Examining Table 5.2, we have found that the net corruption importers and the net corruption exporters belong to developed and developing countries, respectively. This result suggests generally that developed countries are in a position to help developing countries combat corruption. For example, if a lot of corruption embodied in clothing flows from Bangladesh to Japan, as a big contractor, Japan can design local policy to restrict imports without *corruption-clean company* logos, which at least could pressure Bangladesh to combat corruption in its clothing industry.

5.3.2 Corruption Footprint Distribution

To further analyse how corruption embodied in international trade contributes to the entire corruption situation in net importer countries, we provide the corruption footprint contribution in Fig. 5.1. As the least corrupt countries, Switzerland, Australia, Norway and Singapore, which are also shown in Table 5.1, show that imported corruption contributes around a half of the entire corruption footprint.

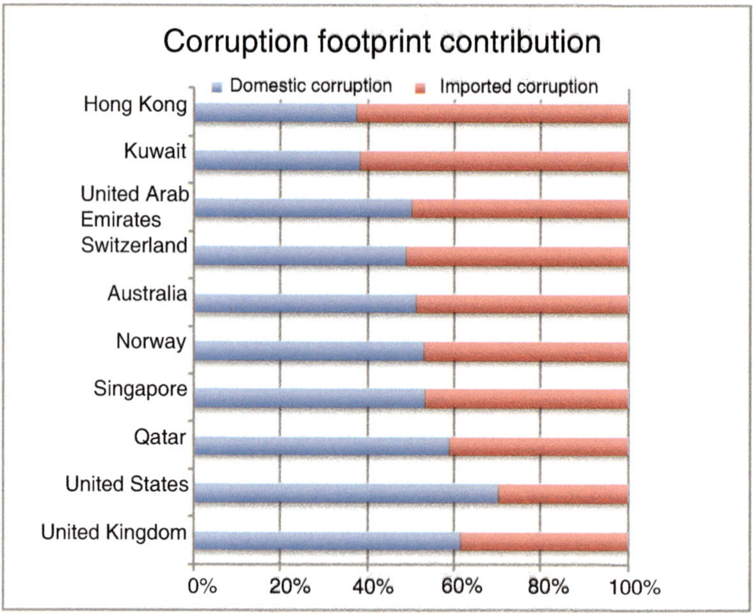

Figure 5.1 Corruption footprint distribution for top 10 net importers.

Assuming that there is an awareness of the fact that corruption embodied in international trade flows to corruption-clean countries, of course the corruption-clean countries could adjust their trading strategies to find more clean partners to decrease the social cost of international trade. More importantly, the clean countries could help the current corrupt partners to combat corruption along the supply chain. For example, relative to its biggest trading partner China, Hong Kong has a well-developed regulatory system to avoid corruption domestically. When importing goods and services from China's corrupt industries, Hong Kong could pressure China to develop corresponding policies to combat corruption. Besides, China could learn from Hong Kong to apply successful proven experience in managing clean environments. This one-on-one help is like an experienced tutor giving a student a personal, tailored learning opportunity, which could make more efficient and targeted improvements.

5.3.3 Corruption-Embodied Commodity Details

Generally, the consumer plays a dominant role in international trade; thus the consumer could pressure the trading partners to make improvements in anti-corruption strategies and policies. To this end, we provide main trading countries and commodity details for the top 10 net corruption importers in Table 5.3. It is a reference for clean countries, indicating where they can contribute their efforts to help developing countries improve their corrupt environment. On the basis of the details of commodities with embodied corruption, the trading partners could make the anti-corruption campaign more targeted.

As shown in Table 5.3, for all the top 10 net corruption importers, China and India are the two main trading partners. The most corruption-embodied commodities in China are items related to manufacture, and the clothing sector tops the corruption-embodied commodity list. India exports different types of commodities with embodied corruption when trading with different partners. Comparing these two net corruption exporters, India's corruption-embodied commodities are mostly raw materials, while China's exported corruption is embodied in more processed commodities. This major difference can be attributed to the economic structure disparity within these two countries. Even though these two countries have the same corruption-embodied commodity, clothing, corruption exists in different guises within the same category.

Table 5.3 Net importers' main trading countries and commodity details

		Main imports (import origin, % of total corruption)		
1	Hong Kong	CHN,[a] 73.9	IND,[b] 3.1	TZA,[c] 2.5
		Clothing, crochet fabric, textiles	Leather products, meat, clothing	Animal skins, fish flours, cut flowers
2	Kuwait	IND, 55.9	CHN, 11.6	TZA, 4
		Rice, raw sugar, cane, clothing	Clothing, textiles, linseed oil, crude	Dried legumes, fish flours, prepared or preserved fish
3	United Arab Emirates	IND, 52.8	CHN, 18.4	TZA, 4.8
		Bread, pastry and cakes, textiles, clothing	Textiles, clothing, floating platforms	Coffee, not roasted; vegetable saps and extracts; fish fillet or meat
4	Switzerland	DEU,[d] 13.7	CHN, 12.8	IND, 12.3
		Bread, pastry and cakes, tourism, clothing	Footwear, automatic data processing machines, leather products	Diamonds; oxygen-function amino-compounds; coffee, not roasted
5	Australia	CHN, 36.3	VNM,[e] 8.6	IND, 8
		Automatic data processing machines; transmission apparatus for radio, telephone, and TV; furniture parts	Tourism, motorbikes, personal household appliances	Diamond, jewelry of precious metal, tea

		Main imports (import origin, % of total corruption)		
6	Norway	CHN, 23.6	DEU, 8.5	IND, 8.4
		Clothing, tourism, leather product	Cars, furniture parts, wine of fresh grapes	Clothing, malt extract, miscellaneous nuts
7	Singapore	IDN,[f] 22.8	CHN, 21.4	IND, 12.8
		Communication equipment, clothing, petroleum gases	Toys and recreation products, clothing, communication equipment	Petroleum oils, refined; extracts of coffee, tea, or mate; clothing
8	Qatar	IND, 32	CHN, 16.6	PAK,[g] 4.2
		Cigars; raw sugar, cane; petroleum oils, refined	Clothing, leather products, textiles	Textiles, rice, cars
9	America	CHN, 37.9	IND, 10.2	MEX,[h] 8.3
		Clothing, leather products, toys and recreation products	Petroleum oils, refined; clothing; medicaments, packaged	Cars, motor vehicle parts, clothing
10	United Kingdom	CHN, 22.9	IND, 13.7	DEU, 6.4
		Clothing, leather product, toys, and recreation products	Tourism, clothing, rice	Clothing, cars, bread, pastry and cakes

*This table is an excerpt (top 10 countries in rows and top 3 trading countries in columns) from Table S4 in Ref. [12].

[a] CHN: China

[b] IND: India

[c] TZA: Tanzania

[d] DEU: Germany

[e] VNM: Vietnam

[f] IDN: Indonesia

[g] PAK: Pakistan

[h] MEX: Mexico

China produces miscellaneous clothing, including top-, middle- and low-priced products. Clothing corruption comes about through unreasonable deductions from funds that should be used in reaching standard working conditions for poor clothing workers, irregular procurement of raw materials and gift-giving of luxury brand clothing to cement social and business networks [13, 14]. It seems China does not have enough comprehensive regulations to regulate business behaviour to maintain a clean environment. Therefore, China could develop anti-corruption policies in the clothing industry to ameliorate corruption along the clothing supply chain. However, India's clothing corruption originates from various license and permit requirements in business management [15, 16]. Corruption has historical nationwide roots in India, so a specific policy in managing the clothing industry might be of little use to change the current situation. In other words, India needs anti-corruption campaigns based on a more integrated view.

5.4 Conclusion

Corruption, as a social indicator in analysing supply chain effects of international trade, can be a supplement to various environmental, economic and other social indicators for triple-bottom-line analysis. In this chapter the corruption embodied in international trade among 189 countries is discussed. Generally, the net corruption importers are developed countries, while the net corruption exporters are developing countries. This finding suggests that developed countries can help developing countries to combat corruption through the vehicle of international trade. Our results also provide the net corruption importers' trading countries and commodity details, which can be a reference for community groups, NGOs and governments to develop targeted anti-corruption policies.

References

1. De Jong, E. and Bogmans, C. (2011). Does corruption discourage international trade?, *European Journal of Political Economy*, **27**(2), 385–398.

2. Krueger, A. O. (1974). The political economy of the rent-seeking society, *The American Economic Review*, **64**(3), 291–303.
3. Ades, A. and Di Tella, R. (1999). Rents, competition, and corruption, *The American Economic Review*, **89**, 982–993.
4. Wei, S.-J. and Shleifer, A. (2000). Local corruption and global capital flows, *Brookings Papers on Economic Activity*, **31**(2), 303–346.
5. Gatti, R. (2004). Explaining corruption: are open countries less corrupt?, *Journal of International Development*, **16**(6), 851–861.
6. Knorich, J. (2004). *International Trade and Corruption: The Influence of Trading Partners*, Seton Hall University dissertations and theses (ETDs), Paper 682.
7. Benoît Norris, C. (2014). Data for social LCA, *International Journal of Life Cycle Assessment*, **19**(2), 261–265.
8. Benoît Norris, C., Norris, G. A. and Aulisio, D. (2014). Efficient assessment of social hotspots in the supply chains of 100 product categories using the social hotspots database, *Sustainability*, **6**(10), 6973–6984.
9. Transparency International. (1995). *TI Corruption Index*, https://www.transparency.org/files/content/tool/1995_CPI_EN.pdf.
10. The World Bank. (2015). *Worldwide Governance Indicators*, http://info.worldbank.org/governance/wgi/index.aspx#doc
11. World Economic Forum. (2011). *The Global Competitiveness Report 2010–2011*, http://www3.weforum.org/docs/WEF_GlobalCompetitivenessReport_2010-11.pdf.
12. Xiao, Y., Lenzen, M., Benoît Norris, C., Norris, G. A., Murray, J. and Malik, A. (2017). The corruption footprints of nations, *Journal of Industrial Ecology*, doi:10.1111/jiec.12537.
13. Business Insider. (2015). *If You're Doing Business in China, Here's Where You'll Find Most Corruption*, http://www.businessinsider.com.au/china-corruption-in-business-study-2015-1.
14. Fashionbi. (2013). *Anti-Gifting Policy in China: How Are Luxury Brands Affected?*, http://fashionbi.com/insights/marketing-analysis/anti-gifting-policy-in-china-how-are-luxury-brands-affected.
15. Deloitte. (2014). *Managing Corruption Risks in India*, http://www2.deloitte.com/content/dam/Deloitte/us/Documents/finance/us-fas-managingcorruptionriskInIndia-05162014.PDF.
16. Xu, B. (2014). *Governance in India: Corruption. Council on Foreign Relations*, http://www.cfr.org/corruption-and-bribery/governance-india-corruption/p31823.

Chapter 6

Social Footprints of Nations: A Look at Welfare

Rachel C. Reyes
Integrated Sustainability Analysis, School of Physics, University of Sydney, Physics Road A28, Sydney, NSW 2006, Australia
rachel.reyes@sydney.edu.au

6.1 Introduction

Globalisation is often seen as having a two-pronged impact on the intertwined domains of the economy, environment and society. It is said to bring about both positive and negative effects: on the one hand are the efficiency and economic gains, but on the other hand are the negative environmental externalities and social costs like persistence of poverty and exacerbation of inequality. The economic impact and environmental repercussions of trade have been well researched and documented, but the social implications remain an area that needs further exploration.

Does the world become better with trade? This chapter presents an attempt at looking at the implications of global trade in terms

The Social Effects of Global Trade
Edited by Joy Murray, Arunima Malik, and Arne Geschke

ISBN 978-981-4774-55-0 (Hardcover), 978-020-3728-71-0 (eBook)
www.panstanford.com

of a summary indicator of the state of society, particularly welfare.[a] Applying multi-region input-output (MRIO) analysis using a social welfare satellite account, embodied welfare in global trade is assessed to answer the question of how much social welfare is associated with our consumption of goods and services. The footprint calculation, originally conceived to quantify the human impact on the ecosystem, is recast to assess the human impact on society. This work complements the existing literature on the application of input-output (IO) analysis that typically uncovers social costs hidden in global supply chains, the likes of the work by Benoît Norris et al. [1], Simas et al. [2] and Alsamawi et al. [3, 4], by showing the scenario through a different lens, that is, the implications of trade for the benefits side, like social progress. In assessing the impact of activities, a complementary measure may prove to be useful. Any measure from one point of view, say merely cost point of view, may be insufficient in presenting an overall picture. Here, to the suite of social footprints, welfare is added.

This work addresses the increasing need for quantifying social impacts by performing social accounting. Indeed, development of social indicators is a prerequisite in understanding and addressing social issues as they relate to economic growth and development.

The rest of this chapter is organised as follows. The next section discusses the initiatives to assess a country's performance beyond the gross domestic product (GDP). Section 6.3 shows the methodology of calculating footprints for a measure that has traditionally not been used for IO calculation, that is, social welfare. Section 6.4 presents the findings of this case study and how from an alternative perspective, a picture of inequality still creeps in in connection with global trade. The last section concludes with remarks on a possible track for sustainability research.

6.2 Beyond the GDP

The central goal of almost all economies is to boost economic activities to generate more income. High growth in the GDP has been

[a]The definition of welfare adopted in this chapter is consistent with the social science definition of welfare as the well-being of individuals and society linked with the quality of life.

the gauge of a country's success since the post-war era. However, the GDP only counts the market value of final goods and services produced within a domestic territory for a particular period. Thus it measures the total expenditure on domestically produced goods, which must coincide with the total income generated by the factors of production employed in producing those goods. Because of income's link to consumption and the assumption that higher consumption implies improvement in the standard of living or the quality of life, the GDP has conventionally been the metric used for assessing life's quality.

How about the other dimensions missed by the GDP but that contribute to overall welfare or social progress? Well-being, for instance, considers health, education, employment status, political voice and governance, social connections and relationships, the environment and the security dimension, among others, besides material living standards. The literature abounds with criticisms of the oversimplification and inappropriateness of merely using the GDP in measuring genuine progress [5–7].

Indeed there are more superior measures of well-being and progress than the GDP. Some of the alternative measures are listed by Costanza et al. [5], such as the Index of Sustainable Economic Welfare (ISEW) and the Genuine Progress Indicator (GPI), which modify the GDP, weighing personal consumption expenditure by income distribution, adding volunteer and household work and subtracting environmental and social costs; the Genuine Savings and Inclusive Wealth Indices, which modify income and capital accounts, respectively; the survey-based indices, such as the Australian Unity Well-Being Index, the World Values Survey, the Gallup-Healthways Well-Being Index and the Gross National Happiness, which look at various dimensions of life satisfaction and the quality of life; and the composite indices, such as the Human Development Index (HDI), the Happy Planet Index (HPI), the Canadian Index of Wellbeing, the National Well-Being Index, the Organisation for Economic Cooperation and Development (OECD) Better Life Index, the Well-Being of Nations and the Sustainable Society Index. This latter set of composite indices have been developed for a range of 36 to 180 countries and investigate indicators on income, health, education, subjective well-being, environment, community engagement, living standards, time use, work-life balance, and human, social and natural capital, among others. Another recent addition to the suite of well-being metrics is the Social Progress Index [8], which takes the GDP totally out of the equation and utilises exclusively social and environmental indicators. See Box 6.1 for more information on the Social Progress Index framework.

Box 6.1 Social Progress Index framework.

Source: Adopted from Social Progress Imperative [8].

The Social Progress Index is a composite indicator designed as a distinctly non-economic measure of social performance that excludes the GDP and instead focuses on outcomes. It synthesises multiple outcome measures that fall under the three broad dimensions of basic human needs, foundations of well-being and opportunity, via a weighting system based on principal component factor analysis. The figure in this box shows the sub-components of these dimensions.

The initiative has shown how social progress relates to per capita GDP and finds a positive but non-linear relationship, in general.

6.3 The Social Welfare Footprint

Footprint exercises have typically accounted for the negative externalities and social costs of consuming goods and services produced in the global production environment, essentially assessing the negative implications of trade. The alternative perspective offered here is a look at the social progress that comes hand in hand with linkages in the globalised world, that is, the social welfare footprint. To complement existing social assessments, this work highlights the positive outcome of trade and traces positive impact throughout the supply chain.

The social welfare footprint is calculated following the standard IO arithmetic. Starting with the IO identity in matrix format $\mathbf{x} = \mathbf{T1} + \mathbf{y}$, read as total production or output, $\mathbf{x}$ is the sum of intermediate demand $\mathbf{T}$, that is, the inter-industry transactions in the economy, and final demand $\mathbf{y}$. The transactions matrix is then expressed as a product of per unit of output direct requirements or input coefficients matrix $\mathbf{A}$ and total output $\mathbf{x}$ so that the IO identity becomes $\mathbf{x} = \mathbf{Ax} + \mathbf{y}$. This allows arriving at the solution to the system as $\mathbf{x} = (\mathbf{I} - \mathbf{A})^{-1}\,\mathbf{y}$, alternatively expressed as $\mathbf{x} = \mathbf{Ly}$, where $\mathbf{I}$ is a conformable identity matrix and $\mathbf{L} = (\mathbf{I} - \mathbf{A})^{-1}$ is the well-known Leontief inverse. While the input coefficients matrix $\mathbf{A}$ only describes the direct links between industries, the Leontief inverse $\mathbf{L}$ captures both the direct and the indirect input requirements for meeting a dollar of final demand.

Introducing a social welfare satellite account $\mathbf{Q}$, expressed per unit of output $\mathbf{q}$, allows the generation of the so-called $\mathbf{m}$ multipliers, which capture direct and indirect social impacts spanning throughout the inter-industry linkage of the supply chain, and are allocated by the Leontief inverse $\mathbf{L}$, via $\mathbf{qx} = \mathbf{q}(\mathbf{I} - \mathbf{A})^{-1}\,\mathbf{y} = \mathbf{my}$. These indicator values account for both direct and embodied values of the social welfare satellite account referenced against a dollar of final demand or per dollar spent on everyday consumption, otherwise known as footprints.

The satellite account in any footprint calculation should have the same level of sectoral disaggregation as the IO structure used in the analysis, in this case the Eora MRIO, which has a total of 14,839 productive sectors for 189 countries [9, 10]. But the developed well-being and progress indicators listed in the previous section are all national-level metrics, thus posing a challenge in establishing connections between an overall measure of social welfare and the supply chain. Moreover, the social welfare values are indices, while satellite accounts in the IO calculation are necessarily in physical units in order to be additive.

This study addresses the above issues by converting the national welfare index into sectoral employment units before performing the consumption-based IO footprint calculation. This means that aside from the social welfare index, details of employment per sector are needed to allocate welfare to workers. This exercise is performed with the assumption that workers who offer their labour service for production of goods and services that flow globally throughout the supply chain experience, on average, the same state of welfare as that represented by their corresponding national index value regardless of the sector they work in. This is admittedly a restrictive assumption, but until social indicators at the sectoral level are tracked, this weighting scheme is a reasonable option on how to operationalise the analysis.

However, the resulting footprints computed this way will be expressed in labour units and not welfare units. To revert back to the associated welfare index embodied in the final consumption, an employment footprint calculation is also performed and the results are used for normalising the employment-weighted social welfare index with employment footprints so that the final indicator is solely in terms of the welfare index. A comparison of social welfare with the economic welfare of workers or income footprints associated with trade is also done.

In this case study, the Social Progress Index is selected as the social welfare indicator because of the comprehensiveness of dimensions it covers. It is also devoid of the economic dimension that allows further analysis of results in comparison with the economic aspect of trade. The employment satellite in terms of full-time equivalent (FTE) units based on the International Labour Organization (ILO) [3] and the income satellite referring to the purchasing power parity (PPP)-adjusted compensation of workers [11] are likewise used.

6.4 Findings

6.4.1 Embodied Welfare in Consumption

Embodied in the goods and services people consume are the inputs that go into the production of those final products and services, including the labour input. The labour input provided by workers is compensated with salaries and wages, and this labour income allows workers to enjoy a particular quality of life related with material well-being. At the same time, these workers experience a certain degree of social welfare, as demonstrated by their ability to satisfy basic human needs, the quality of their health and ecosystem and the opportunities they can access dependent on the structure of the environment they revolve in. In the absence of detailed accounting, it is assumed that each worker in a particular country enjoys a uniform state of social welfare captured in the national welfare index.

It can be gleaned from the foregoing that the economic structure supported by people's consumption behaviour is associated with a quality of life enjoyed by workers participating in that system. This link allows us to draw worker welfare implications concomitant with consumption. Although the link does not capture causality and no claim is made that a change in consumption of a set of agents[b] causes a change in welfare in another set, it establishes an association so that it can be said that consumption may have welfare implications beyond the domestic territory of buyers. For example, if a buyer in country A buys goods imported from country B, country A's consumption may have welfare implications for workers of country B. A calculation following this logic yields the results shown in Figs. 6.1 and 6.2 for the top net exporters and net importers of embodied social progress, respectively.

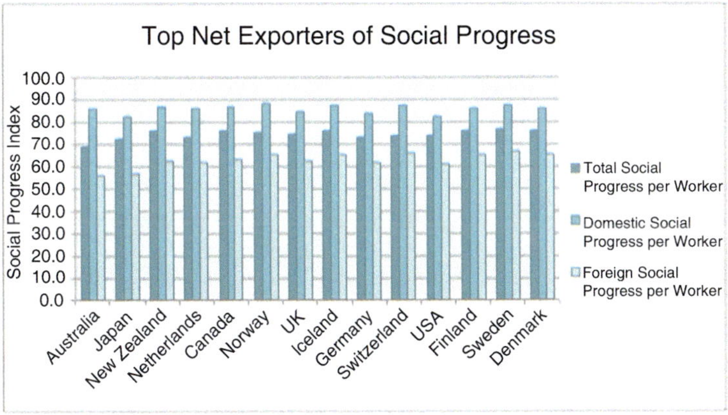

Figure 6.1 Countries with the largest net exports of embodied social progress and their associated social footprint values.

The results presented show three footprint measures: the total social progress per worker, the domestic social progress per worker and the foreign, or imported, social progress per worker. The *total social progress footprint* captures the social welfare index supported by a country's consumption regardless of the origin of goods, that

[b]Agents refer to economic players and in this particular case consumers in a certain geographic area.

is, with no distinction between the domestic and imported sources. The *domestic social progress footprint* refers to the social welfare index associated with purchases of goods produced by workers in their own countries, thus equivalent to the original national Social Progress Index. Finally, the *foreign social progress footprint* refers to the social welfare index associated with imported goods, or those produced by foreign workers, and therefore captures the implications of domestic consumption for the well-being of foreign workers.

Figure 6.1 reveals the developed countries of Australia, Japan, New Zealand, the Netherlands, Canada, Norway, the United Kingdom, Iceland, Germany, Switzerland, the United States, Finland, Sweden and Denmark as net exporters of social progress. The consumption of these rich countries is associated with relatively higher domestic social welfare, while the social welfare embodied in their imports is much lower.

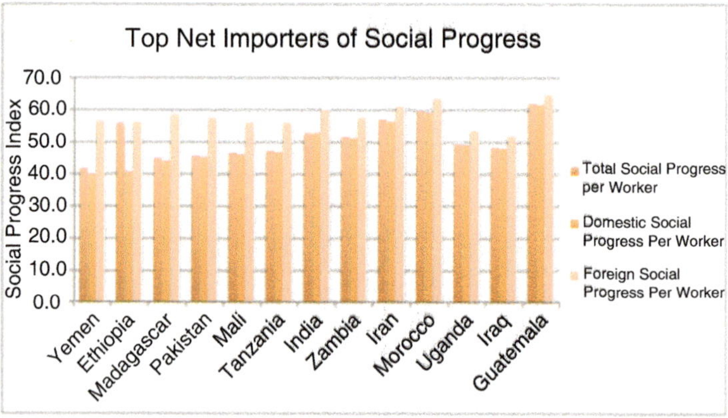

Figure 6.2 Countries with the largest net imports of embodied social progress and their associated social footprint values.

On the other hand, Fig. 6.2 depicts the opposite picture. The developing economies of Yemen, Ethiopia, Madagascar, Pakistan, Mali, Tanzania, India, Zambia, Iran, Morocco, Uganda, Iraq and Guatemala are shown to be net importers of social progress. This implies that despite their relatively low domestic social welfare, their consumption appears to coincide with higher overall welfare because the imported component of their purchases embodies higher worker social welfare.

6.4.2 Inequality

Despite the switch of perspective to accounting for positive implications of trade instead of negative externalities or social costs by the likes of the work by Benoît Norris et al. [1], Simas et al. [2] and Alsamawi et al. [3, 4], a significant social issue is still uncovered. Putting together the top-ranking and bottom-ranking countries in terms of the ratio of domestic to foreign social welfare footprint, Fig. 6.3 exposes the wide divide or inequality of worker welfare in rich and poor countries. Notice how domestic workers are better off than their foreign counterparts in rich countries, as demonstrated by ratios greater than 1, while the story is reversed for poor countries, with ratios under 1.

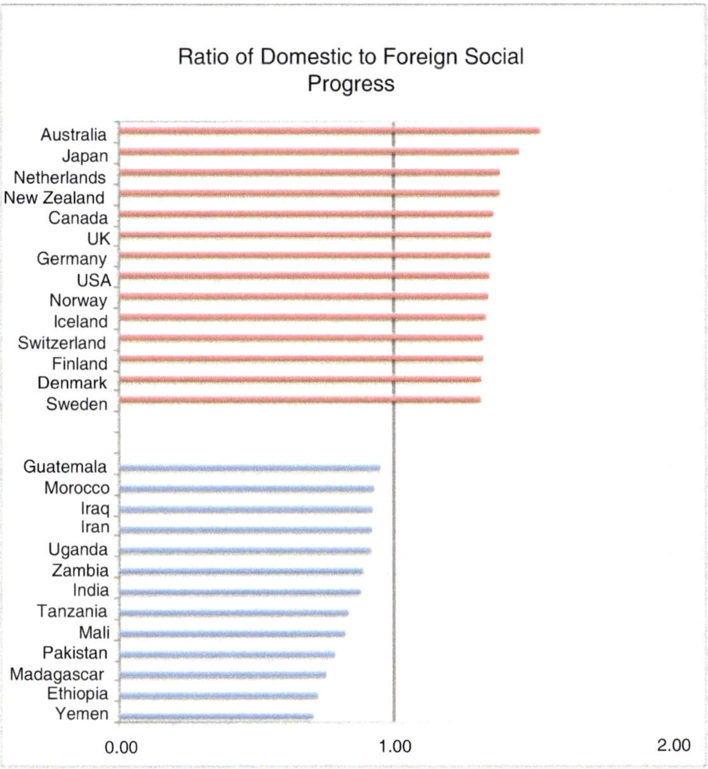

Figure 6.3 Ratio of domestic social footprint to foreign social footprint for the top net exporter and top net importer countries. The upper panel comprises the net exporter countries with ratios over 1, while the lower panel comprises the net importer countries with ratios under 1.

6.4.3 Social Welfare and Economic Welfare

Finally, putting together the footprints capturing social welfare (social progress footprints), with the footprints representing economic or material welfare of workers (PPP-adjusted income footprints), a positive correlation is established. There tends to be a generally positive non-linear relationship between the two welfare indicators. As average wage increases, social progress also appears to increase but at a decreasing rate (Fig. 6.4). This suggests that higher income tends to be associated with better well-being, but after breaching a certain level of income, well-being does not necessarily become higher, as shown by the flattening of the fitted curve. A very high income level could in fact even be associated with lower well-being. This is consistent with findings on the association between economic and non-economic measures of welfare in the literature.

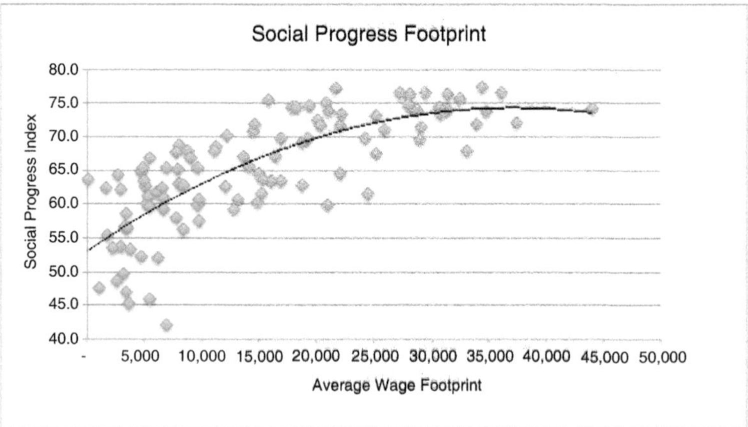

Figure 6.4 Chart showing the association between social welfare in terms of the social progress footprint and economic welfare in terms of the average-wage-of-workers' footprint.

6.5 Going Forward

The procedure adopted in this case study is an initial attempt to establish an association between a welfare summary statistic and the supply chain to trace embodied social welfare in consumption.

The technique is far from ideal, yet the analysis provides consistent results that can potentially be of aid in informing policy. The findings in this case study complement work that has been done to uncover issues hidden in the complex supply chains.

Going forward, development of sector-specific welfare indicators may shed more light on social issues and drastically improve the richness of findings and implications arising from such analysis. Meanwhile, this work illustrates the possibility of performing a social footprint analysis covering overall welfare.

Acknowledgements

The author wishes to thank Prof. Manfred Lenzen for his contribution and Dr. Joy Murray for the insights she shared on this topic.

References

1. Benoît Norris, C., Aulisio Cavan, D. and Norris, G. (2012). Identifying social impacts in product supply chains: overview and application of the Social Hotspots Database, *Sustainability*, **4**, 1946–1965.
2. Simas, M. S., Golsteijn, L., Huijbregts, M. A. J., Wood, R. and Hertwich, E. G. (2014). The 'bad labor' footprint: quantifying the social impacts of globalization, *Sustainability*, **6**, 7514–7540.
3. Alsamawi, A., Murray, J. and Lenzen, M. (2014). The employment footprints of nations: uncovering master-servant relationships, *Journal of Industrial Ecology*, **18**(1), 59–70.
4. Alsamawi, A., Murray, J., Lenzen, M., Moran, D. and Kanemoto, K. (2014). The inequality footprints of nations: a novel approach to quantitative accounting of income inequality, *PLOS ONE*, **9**(10), e110881.
5. Costanza, R., Kubiszewski, I., Giovannini, E., Lovins, H., McGlade, J., Pickett, K. E., Vala Ragnarsdóttir, K., Roberts, D., De Vogli, R. and Wilkinson, R. (2014). Time to leave GDP behind, *Nature*, **505**, 283–285.
6. Kubiszewski, I., Costanza, R., Franco, C., Lawn, P., Talberth, J., Jackson, T. and Aylmer, C. (2013). Beyond GDP: measuring and achieving global genuine progress, *Ecological Economics*, **93**, 57–68.
7. Victor, P. (2010). Questioning economic growth, *Nature*, **468**, 370–371.
8. Social Progress Imperative. (2015). *Social Progress Index 2015 Report*, http://www.socialprogressimperative.org.

9. Lenzen, M., Kanemoto, K., Moran, D. and Geschke, A. (2012). Mapping the structure of the world economy, *Environmental Science & Technology*, **46**(15), 8374–8381.
10. Lenzen, M., Moran, D., Kanemoto, K. and Geschke, A. (2013). Building Eora: a multi-region input-output database at high country and sector resolution, *Economic Systems Research*, **25**(1), 20–49.
11. Reyes, R. C., Lenzen, M. and Murray, J. (2016). Better global assessment of worker inequality: comment on 'the employment footprints of nations', *Journal of Industrial Ecology*, doi:10.1111/jiec.12489.

Chapter 7

Employment Flows from and into the Arab Region: A Case Study to Measure the Embodied Employment in 2010

Ali Alsamawi[a] and Yanyan Xiao[b]
[a]*The Organisation for Economic Co-operation and Development, 2 rue André Pascal, 75775 Paris, Cedex 16, France*
[b]*ISA, School of Physics A28, University of Sydney, Sydney, NSW 2006, Australia*
ali.alsamawi@oecd.org; yanyan.xiao@sydney.edu.au

7.1 Introduction

Countries import goods and services to satisfy their domestic needs, and those goods and services, in turn, require employment and income payments in the producing countries [1]. Following evidence of the appalling working conditions of some of those workers in the producing countries (e.g., the collapse of the Rana Plaza building in Bangladesh [2]) enterprises, governments and international organisations are now working to improve the conditions of workers in their supply chains. For instance, the Dutch government recently

The Social Effects of Global Trade
Edited by Joy Murray, Arunima Malik, and Arne Geschke

ISBN 978-981-4774-55-0 (Hardcover), 978-020-3728-71-0 (eBook)
www.panstanford.com

planned to ban companies that involve children in the production of imported products.

Inter-relationships between international trade and social issues have been considerably neglected in past decades [3]. The amount of cross-border goods and services has increased significantly, reaching around US$16 trillion in 2008 [4]. This huge amount of international trade reflects the number of workers worldwide working to produce these cross-border goods and services. According to Ref. [5], 20% of the world's workforce in 2010 was working in export-oriented sectors and hence its labour was embodied in global supply chains (this 20% is equivalent to about 500 million workers).

After successfully applying a global multi-region input-output (MRIO) model to evaluate environmental issues (e.g., carbon footprints [6] and virtual water [7]), we recognise that this technique can also be used to examine employment, occupational health and safety, poverty, and other social issues [5, 8, 9].

7.2 Methodology

This study uses prior work from Ref. [5] with a focus on the Arab region to discuss employment footprints. The employment footprint of a nation includes the number of workers working for the domestic market plus the number of workers embodied in its traded imports (this will be discussed later in the chapter). We examine this region because we expect to find that a large number of workers support the demand of the Arab countries and, in particular, the Gulf Cooperation Countries (GCC) through traded products. While the GCC may benefit from a net inflow of embodied labour in imported products, other parts of the Arab region, in particular the Maghreb-region countries, heavily rely on producing and exporting products that are consumed in Europe, leading to a net outflow of embodied labour. Quantifying the flow of labour will help those Arab countries to identify how much employment each country requires in addition to its own domestic workforce. The outcome of our research provides information on the number of workers within a country required to produce products demanded by other countries (workers working in export-oriented industry).

In Ref. [5], we define employment footprints as 'a country's domestic employment and that recruited along the supply chains of, and hence embodied in, its imported goods and services' (where 'domestic' refers to the workers working for the domestic market).

These footprints reflect the full-time equivalent (FTE) employment required to satisfy a given country's consumption directly (through the domestic workforce) and indirectly (through embodied labour within imported products). The methodology used here is the same as that used in the carbon footprint studies that rely on an economic MRIO table, a physical satellite account, and Leontief's demand-pull impact calculation (in this case employment; see Ref. [5] for more information about the methodology).

7.3 Results

In the Arab region, there is a significant difference between countries in terms of flow of embodied labour (Fig. 7.1). Countries with high gross domestic products (GDPs) have more embodied labour associated with imported products than countries with low GDPs. In this regard, Saudi Arabia (41% of total Arab intra-regional flows) and the United Arab Emirates (10%) jointly imported half of the embodied employment that flowed from other Arab countries in 2010. On the other hand, Sudan (27% of total Arab intra-regional employment flows) and Egypt (21%) jointly constituted about half of the total employment export flows to Arab countries (see Figs. 7.1 and 7.2).

In addition, Fig. 7.1 shows that the flows of embodied labour from Maghreb countries (e.g., Algeria and Morocco) to other Arab sub-regions are the lowest (we have not shown the flow of embodied labour from Algeria in the figure because the economic linkages between Algeria and other Arab countries are very low). This is perhaps because (i) there are some trade treaties between Arab countries and European Union (EU) countries and, thus, a large amount of exports of agricultural products, food manufacturing, garment and others flow from Arab countries to EU countries [10]; and (ii) according to the gravity model of trade, the trade between two countries can be affected by the distance between them. It can be affected, for instance, by the costs of transportation and insurance on the exported and imported goods.

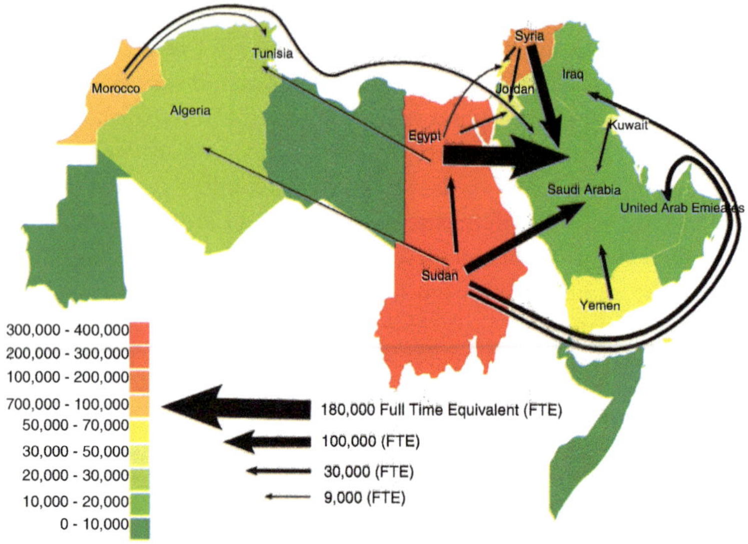

Figure 7.1 Top flows of embodied labour between Arab countries in 2010. The colours reflect the level of total embodied labour in the exports of Arab countries into other Arab countries, and the arrows reflect the direction of bilateral flows.

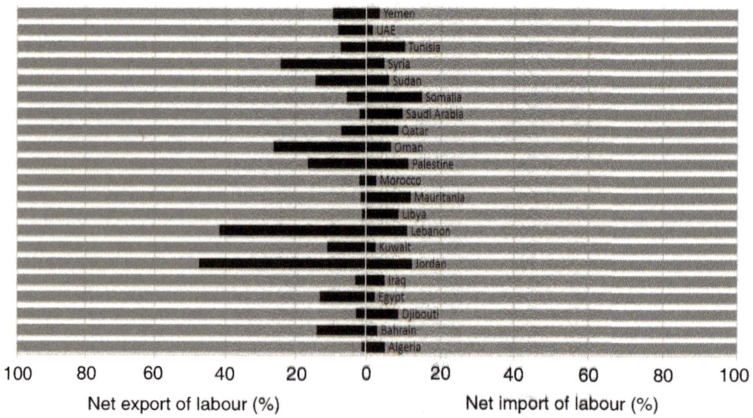

Figure 7.2 Export and import of embodied labour from Arab countries to other Arab countries (black bars) and the rest of the world (grey bars) as a percentage of the total workforce.

Moreover, there is a difference between the flows of embodied labour from Arab countries to the rest of the world (ROW) and intra-Arab flows (Fig. 7.2 and Table 7.1). In Table 7.1, countries are classified as net exporters and net importers of labour.

Figure 7.2 shows the import and export of labour shares between Arab countries and the ROW. Among high-GDP countries (i.e., GCC), the goods and services consumed by Saudi Arabia include only 10% of embodied labour (about 516,000 FTE workers) from Arab countries and 90% from the ROW, while goods and services consumed by the United Arab Emirates embody only 2% of labour (about 122,000 FTE workers) from Arab countries and 98% from the ROW (Fig. 7.2). Among middle-and low-GDP countries (or Mashreq[a]), Jordan and Lebanon have the largest exported flows to Arab countries, 48% and 42%, respectively, of the total exported labour (Fig. 7.2). Somalia has the highest imported flows of embodied labour from Arab countries (15% of its total embodied labour; Fig. 7.2). This could be attributed to its scarce domestic agriculture resources and, therefore, its reliance on large imports of vegetables, raw sugar, rice and bovine and wheat flours, which are produced by labour-intensive industries.

Table 7.1 compares every country's import and export of labour with its domestic labour force (or workforce), which illustrates the reliance of an economy on imported or exported labour. In the case of Algeria, only 0.2% of its labour force (for a definition, see the note below Table 7.1) is embodied in exports to the Arab countries, while 14.6% went to the ROW (Table 7.1). In contrast, embodied labour from the Arab countries to the United Arab Emirates is equivalent to 6.7% of its workforce while that embodied from the ROW is about four times the United Arab Emirates' workforce (Fig. 7.2 and Table 7.1). This is probably for two reasons: One is that the imported goods are more labour-intensive than the exported goods, and the other is that the very low tariffs and few non-tariff barriers to trade in the United Arab Emirates promote prosperous bilateral trade. Table 7.1 also suggests the unemployment situation in some countries could be an issue. As mentioned before, the big gap between Oman's total imports of labour from the ROW in the labour force (48.4) and its total imports of labour from the ROW in the workforce (208.8) suggests large unemployment there.

[a]Lebanon, Palestine, Jordan and Iraq.

Table 7.1 Amounts of embodied labour in exports and imports as a percentage of the total labour force for Arab countries in 2010 (no data for Comoros)

Country	Total exports to Arab nations (% of labour force[a])	Total exports to ROW[b] (% of labour force)	Total exports to Arab nations (% of workforce)	Total exports to ROW (% of workforce)	Total imports from Arab nations (% of labour force)	Total imports from ROW (% of labour force)	Total imports from Arab nations (% of workforce)	Total imports from ROW (% of workforce)
Algeria	0.2	14.6	0.3	21.6	0.5	9.1	0.7	13.4
Bahrain	1.4	8.4	3.4	20.4	1.1	35.3	2.7	85.6
Djibouti	0.7	23.2	0.7	24.4	1.1	11.2	1.1	11.8
Egypt	1.0	6.8	1.2	8.1	0.3	10.8	0.3	12.9
Iraq	0.2	5.5	0.2	5.7	0.8	15.2	0.8	15.8
Jordan	2.8	3.0	4.4	4.8	6.0	43.0	9.5	68.6
Kuwait	3.1	24.6	4.3	34.1	1.9	72.4	2.7	100.5
Lebanon	4.3	5.9	4.4	6.1	3.2	25.4	3.3	26.3
Libya	0.2	15.1	0.4	26.2	1.0	10.6	1.7	18.5
Mauritania	0.4	21.5	0.4	21.7	0.8	5.7	0.8	5.7
Morocco	0.6	28.9	0.7	32.4	0.1	5.2	0.2	5.9
Oman	1.0	2.8	4.4	12.2	3.4	48.4	14.8	208.8
Palestine	0.7	3.7	1.0	4.9	2.2	17.2	2.9	22.7
Qatar	1.1	14.4	1.8	22.2	3.8	40.5	5.8	62.4

Country	Total exports to Arab nations (% of labour force[a])	Total exports to ROW[b] (% of labour force)	Total exports to Arab nations (% of workforce)	Total exports to ROW (% of workforce)	Total imports from Arab nations (% of labour force)	Total imports from ROW (% of labour force)	Total imports from Arab nations (% of workforce)	Total imports from ROW (% of workforce)
Saudi Arabia	0.1	6.8	0.2	8.3	5.3	49.4	6.5	60.3
Somalia	0.1	1.3	0.1	1.3	0.1	0.7	0.1	0.7
Sudan	8.1	15.3	11.1	22.5	0.1	1.1	0.1	1.2
Syria	3.5	10.7	4.3	13.2	0.3	6.7	0.4	8.2
Tunisia	0.6	7.2	0.9	10.6	1.5	13.0	2.2	19.2
United Arab Emirates	0.3	2.8	0.8	8.9	2.1	130.8	6.7	407.1
Yemen	1.1	9.7	1.9	17.0	0.1	3.5	0.2	6.1

Source: The World Bank and Ref. [5]

[a] Labour force (all people who are able to work) = Workforce + Unemployment

[b] ROW = Rest of the world

Note: Reading across the top row: The value 0.2% represents the percentage of Algeria's labour force that is working solely for its exports to other Arab nations, 14.6% of the labour force is working solely for its exports to the rest of the world, 0.3% represents the percentage of Algeria's workforce that is working solely for its exports to other Arab nations and 21.6% represents the labour force for its exports to the ROW. The imported labour from other Arab nations is equivalent to 0.5% of Algeria's own labour force and imported labour from the ROW is equivalent to 9.1% of Algeria's own labour force; in the case of the workforce, the total imported labour from other Arab nations is equivalent to 0.7% of Algeria's own workforce and labour imported from the ROW is equivalent to 13.4% of Algeria's workforce. The comparison between labour force and workforce indicates the level of unemployment in an economy. For example, Oman's total labour imports from the ROW (% of labour force) is 48.4, while its total labour imports from the ROW (% of workforce) is 208.8; the big gap suggests large unemployment.

Investigation of sectors linked to exported products from Arab countries to Arab countries reveals that agriculture is at the top, followed by textile products [10]. As mentioned before, Egypt has the highest employment flows to the Arab countries. According to the findings, agriculture products and food products were the major exports in 2010. To produce these products, children are often forced to work to help their families, which takes the form of unpaid work [11]. In the case of Sudan, most of Sudanese exports to the Arab countries are livestock products, which are produced by the sector characterised by poor working conditions [12, 13]. Meanwhile, most of labour flows from GCC countries to other Arab sub-regions come from the oil and gas sectors [10]. Generally, high-GDP countries in the Arab region are mining-dominated countries, while the mining sector has a low workforce, which is why the volume of labour flowing from high-GDP countries is small.

7.4 Conclusion

Saudi Arabia and other GCC countries rely on offshore labour (i.e., embodied in imports from other countries) for providing agricultural products, because their own climate is unsuitable for growing these products. Moreover, water in these nations is scarce. In addition, the booming oil and other mineral industries seek employment from immediate neighbours, which are Arab countries that offer low income to workers.

In a nutshell, labour-abundant countries (such as Egypt and Sudan) export labour-intensive goods (such as agriculture products), while capital-abundant countries (such as Saudi Arabia and the United Arab Emirates) export capital-intensive goods (oil and gas) [14]. These results can be used to determine how much of the workforce from the Arab region and the ROW each Arab country requires, in addition to its national workforce, to satisfy its domestic consumption. Also the results allow a mapping of the integration, which can present an overview of the employment flows from and into the Arab countries.

Robust measurements on jobs created by trade and those generated by the domestic market provide hard data on which to base policy and practices. The data can be used to indicate the benefits of trade. From a consumer perspective, the embodied workers in imports satisfy the importing countries' needs for goods and services. In contrast, from a producer perspective, increasing trade will help the exporting countries to create new jobs. This chapter provides information about the number of workers working in exports and thus can present a better understanding of the economic value created by those workers.

References

1. Kol, J. (1991). Key sectors, comparative advantage, and international shifts in employment: a case study for Indonesia, South Korea, Mexico and Pakistan and their trade relations with the European Community, in *Advances in Input-Output Analysis*, ed. Peterson, W. (Oxford University Press, New York), pp. 199–210.
2. Bangladesh All Party Parliamentary Group. (2013). *After Rana Plaza: A Report into the Readymade Garment Industry in Bangladesh*, Parliamentary, UK Parliament, Parliamentary Liaison Office, London, p. 60, http://www.annemain.com/pdf/APPG_Bangladesh_Garment_Industry_Report.pdf.
3. Becker, J. (2017). The Netherlands's plan to cut child labor out of products. Human Rights Watch news (accessed on 15 March 2017).
4. Krugman, P., Obstfeld, M. and Melitz, M. (2012). *International Economics: Theory and Policy*, 9th ed. (Pearson Addison-Wesley, Boston).
5. Alsamawi, A., Murray, J. and Lenzen, M. (2014). The employment footprints of nations, *Journal of Industrial Ecology*, **18**, 59–70.
6. Hertwich, E. G. and Peters, G. P. (2009). Carbon footprint of nations: a global, trade-linked analysis, *Environmental Science & Technology*, **43**(16), 6414–6420.
7. Feng, K., Chapagain, A. K., Suh, S., Pfister, S. and Hubacek, K. (2011). Comparison of bottom-up and top-down approaches to calculating the water footprints of nations, *Economic Systems Research*, **23**(4), 371–385.
8. Ghosh, A. and Sengupta, A. (1984). *Income Distribution and the Structure of Production in an Input-Output Framework* (United Nations Industrial Development Organization, New York).
9. Groenewold, N., Hagger, A. J. and Madden, J. R. (1987). The measurement of industry employment contribution in an input-output model, *Regional Studies*, **21**, 255–263.

10. Hausmann, R., Hidalgo, C., Bustos, S., Coscia, M., Chung, S., Jiménez, J., Simoes, A. and Yildirim, M. (2011). *The Atlas of Economic Complexity. Mapping Paths to Prosperity* (Harvard University, Center for International Development, MIT Media Lab, Cambridge).
11. International Labour Organization. (2012). *Working Children in Egypt: Results of the 2010 National Child Labour Survey*, International Programme on the Elimination of Child Labour.
12. http://www.antislavery.org/english/slavery_today/forced_labour_in_sudan.aspx (accessed in July 2016).
13. Babiker, I. B., Abdul-Jabbar, M. and Al-Feel, M. A. (2011). Sudanese live sheep and mutton exports competitiveness, *Journal of the Saudi Society of Agricultural Sciences*, **10**, 25–32.
14. Heckscher, E. F. and Ohlin, B. G. (1991). *Heckscher-Ohlin Trade Theory* (MIT Press, Cambridge; London).

Chapter 8

The Distribution of Labour and Wages Embodied in European Consumption

Moana Simas and Richard Wood
Industrial Ecology Program, Norwegian University of Science and Technology, Sem Sælands vei 7, 7034 Trondheim, Norway
moana.s.simas@ntnu.no

8.1 Introduction

The last two decades represented a substantial increase in the volume of products and services traded internationally. Production chains became spread all over the world, leading to a new division of labour, where management, production and resource extraction can happen in substantially different places on earth. In a third of a century, between 1980 and 2014, imports worldwide have increased over five-fold, mainly driven by imports to high-income countries (Fig. 8.1). Industrialised countries have gradually become net importers of environmental pressures, such as greenhouse gas emissions, energy, materials and water [1] and also of labour [2].

The Social Effects of Global Trade
Edited by Joy Murray, Arunima Malik, and Arne Geschke

ISBN 978-981-4774-55-0 (Hardcover), 978-020-3728-71-0 (eBook)
www.panstanford.com

This occurs mainly because these countries' economies have gone through a shift towards a service-based and high-value-added economy, while increasing the imports of low-value-added and labour- and energy-intensive products. Consumption of goods and services per capita, at the same time, has increased substantially, generating the externalisation of many environmental and social impacts to less developed countries. At the same time, globalisation connects markets around the world, generating work and income along the supply chain.

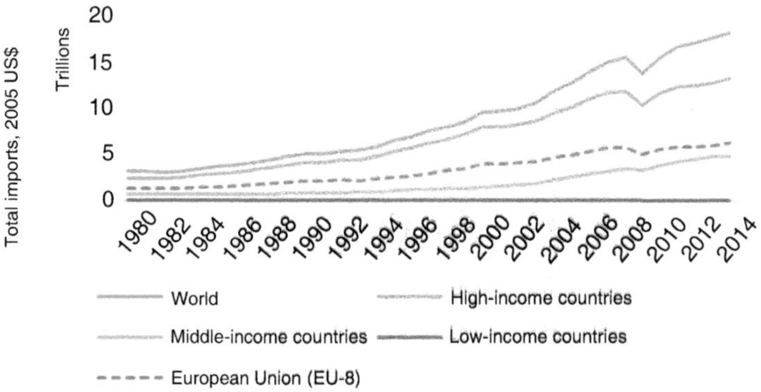

Figure 8.1 Total imports in the world; in the European Union, and in high-, medium- and low-income countries, in trillions 2005 constant US dollars, between 1980 and 2014 [3].

In this study we look at the impacts of European household consumption on socioeconomic impacts generated throughout the globe in the past years. The analysis comprises the period between 1995 and 2011. We use a multi-regional input-output (MRIO) model and EXIOBASE[a] and its socioeconomic extensions to quantify consumption-based footprints for the EU-28 countries. The European Union (EU) demands over one-third of global imports. We look at the origin of impacts of products we consume in our daily life: food products, clothes and electronics. We present an analysis of the origins of workers and salaries embodied in our consumption and where the workers in most vulnerable situations are located.

[a]http://exiobase.eu/

8.2 Accounting for Socioeconomic Impacts in Supply Chains

Consumption-based accounting has been widely used for assessing environmental impacts of consumed products in the form of environmental footprints [4]. Social footprints have just recently started gaining space in the footprint field, with the incorporation of social extensions in environmentally extended economic models and a growing interest in assessing social impacts embodied in international supply chains [5]. In fact, awareness of labour conditions abroad has inspired many campaigns and labels to certify products free of what is seen as undesirable labour conditions. The most famous of these labels is the Fairtrade label, which guarantees fair remuneration and compliance to environmental and labour laws to producers and small traders of agricultural products and textiles. In the same line, the Conflict-Free Sourcing Initiative[b] audits and validates smelters and refiners that supply minerals sourced responsibly in conflict-affected and high-risk areas. The examples from these two initiatives rely on two common characteristics: traceability in supply chains and transparency in supply chains.

The globalisation of supply chains widens the gap between producers and consumers. It creates complexity, which reduces the traceability of conditions in the diverse manufacturing stages of the consumed product. At the same time, globalisation generates jobs and income worldwide [6] and could alleviate poverty in developing countries [7]. Much of social footprinting has focused on negative social aspects of international trade (see, for example, Refs. [8, 9]). However, work and income generation are key indicators to measure socioeconomic development. Improving employment rates is a significant economic goal, and unemployment was ranked as one of the major distresses during the 2007–2008 global financial crisis. Work and income act at the individual level and at the local/regional level. At the individual level work generates income to families and is a part of individual and group identity. At the local and regional levels, work and income drive social and economic progress locally, promoting entrepreneurship and consumption, generating taxes

[b]http://www.conflictfreesourcing.org

and revenues and also improving access for local production to formal and informal markets.

Many regions have increased their level of affluence by becoming exporters of materials and services to other regions and nations. On the other hand, as labour costs are among the main drivers for the relocation of industries to developing countries, it is also important to look at how the wages are distributed throughout the supply chain. The generation of low-income labour abroad is not a bad situation but it raises the question whether globalisation is bringing more benefits to the exporting nations or to the importing nations. The products we consume on a regular basis can be associated with harmful work conditions. Events such as the Rana Plaza building collapse in 2013, which killed over 1000 garment workers involved in the production of clothes for famous brands, or the suicide of 14 workers of Foxxcon protesting work conditions in the manufacture of electronic components and devices in 2010, draw attention to social impacts in production chains. Thus, it becomes necessary to identify not only the quantity and origin of workers and wages in the supply chain of products consumed in the EU but also the whereabouts of the workers in vulnerable work conditions.

8.3 Methods

Consumption-based accounting, or footprinting, consists of allocating impacts that take place throughout the supply chain to the final consumed products. It covers all the production stages of consumed products up to their purchase by final consumers, accounting for extraction of raw materials, processing, transportation and sales. The idea is that impacts are ultimately driven by the final purchase of the products. Thus, energy consumption, waste generation or labour needed for the production and transportation of goods only happens because there is a demand for these products.

Footprinting is performed using an MRIO model, which accounts for production stages in different industries and regions. It tracks the monetary exchange between industries. The MRIO model used in this study is EXIOBASE,[c] which represents the world economy in

[c]You can find the free version 2 of EXIOBASE, which covers the year 2007, at http://www.exiobase.eu/. For details on the methods, refer to Wood et al. [10]. In this study, we used an initial version of EXIOBASE 3, which comprises a time series between 1995 and 2011.

49 regions and 200 products (combined 9800 products), covering a time series from 1995 to 2011. EXIOBASE has a detailed coverage of the 28 countries of the EU,[d] major developed and developing economies, and main EU trade partners, comprising over 85% of world's gross domestic product (GDP) in 2011. The remaining countries are aggregated in five continental regions.

The footprints (f) can be calculated by using the technology and trade matrix (A), the socioeconomic extensions matrix (S), and final demand (y). Footprints are calculated as shown in Eq. 8.1, where I is an identity matrix with the same size as A:

$$f = S(I - A)^{-1} y \tag{8.1}$$

In this study, S is a matrix with coefficients for socioeconomic impacts per unit output for each sector. The socioeconomic extensions included in this assessment are wages in purchasing power parity (PPP), total employment and vulnerable employment. Vulnerable workers are defined as those with no formal work arrangements and more likely to lack decent work condition and include own-account workers and paid or unpaid family workers with no formal employment contracts [9]. Wages were transformed from current euros to constant 2011 international dollars to account for differences in the value of money over time and among different regions. The final demand (y) vector used is the sum of household consumption in the 28 EU countries, and it excludes government purchases, fixed capital formation and change in inventories. As with footprint modelling, it includes all consumption made by residents in a certain area. Thus it includes domestic production consumed in the territory, imports, but excludes exports.

8.4 Socioeconomic Impacts of European Consumption

European household consumption increased by 80% between 1995 and 2011, from 3.2 to 5.8 trillion euros. Figure 8.2 illustrates the growth in household expenditure in the period for food, clothing and electronic products and the total expenditure in constant 1995

---

[d]As at the time of writing, the United Kingdom was formally still a member of the EU, it was, therefore, included as an EU member in this analysis.

prices. The participation of different products in the consumption basket of European families changed in this period.

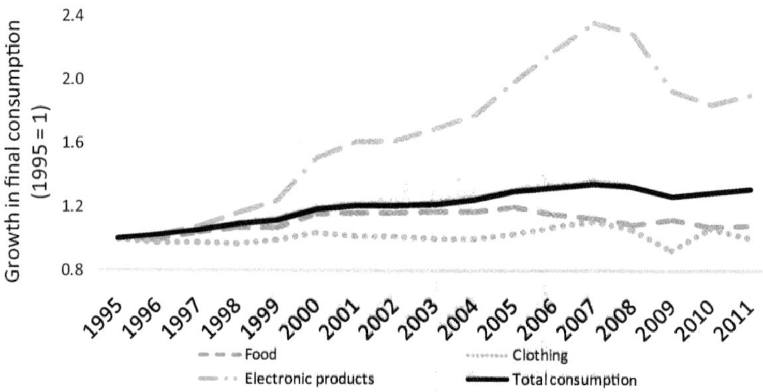

Figure 8.2 Growth in European household consumption between 1995 and 2011, relative to 1995. The filled line represents the household consumption of all goods and services in the EU. The dashed, dotted and dashed-dotted lines represent growth in the household consumption for, respectively, food, clothing and electronic products.

From the late 1990s until 2007, the growth in the consumption of electronic products increased considerably. In 2007, probably because of the financial crisis, the purchase of these goods dropped considerably—by 2009, the expenditure on electronic products had dropped by nearly 20% from the 2007 level, from 232 to 189 billion constant 1995 euros. The total expenditure, in constant 1995 prices, increased by 30% between 1995 and 2011, while spending on food and clothing remained relatively stable throughout the period.

To assess how EU consumption affects workers throughout the world, Fig. 8.3 shows the wages, employment and vulnerable employment footprints of European households. The line plots (Fig. 8.3, top) show the evolution of footprints from 1995 to 2011, and the share of this footprint that happened in Europe (dashed lines) and outside of Europe (dotted lines). The heat maps (Fig. 8.3, bottom) show the distribution of the footprints, in 2011, according to the origin of the impacts, that is, where the wages and workers were situated.

It is noticeable that wages are concentrated in the EU even when controlling for price differences between countries—82%

of wages in current euros and 73% of wages in PPP in 2011 were paid to workers who live within the EU. In contrast, 68% of total employment and 82% of all workers in vulnerable conditions associated with products consumed in Europe took place outside the region. While total employment is distributed mostly between Africa and Asia, workers in vulnerable condition are located mostly in the latter region.

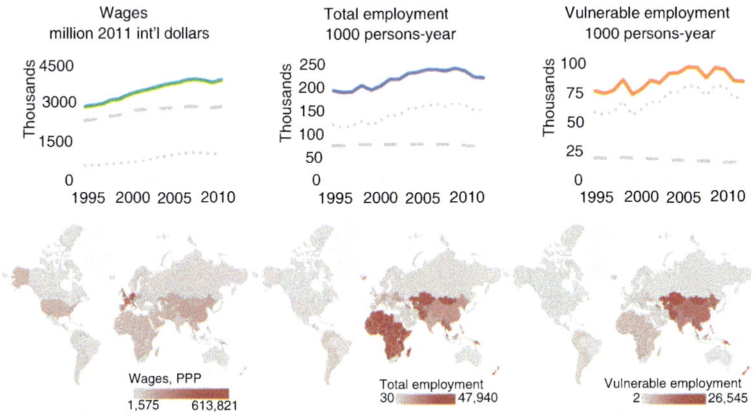

Figure 8.3 Footprints for European total household consumption for wages (left), total employment (centre) and vulnerable employment (right). (Top) Line plots represent the total annual footprint from 1995 to 2011 (coloured, continuous lines), part of the footprint that happened in the European countries (dashed lines) and part of the footprint outside the European Union (dotted lines). (Bottom) The maps show, for each indicator, distribution of impacts across the countries for 2011 footprints. A stronger red colour means a higher concentration of impacts in the country/region.

Wages and employment can be traced to different countries, as well as to different sectors within these countries. Figure 8.4 shows the flows of wages (a), total employment (b) and vulnerable employment (c) for the total European household consumption in 2011. Most of the wages were paid to workers in the services sector in Europe and driven by the consumption of services by European consumers. Most workers, however, were located outside the EU and were associated mainly with the consumption of food products. Workers in agriculture in non-European countries correspond to

39% of the total employment footprint and 58% of the vulnerable employment footprint.

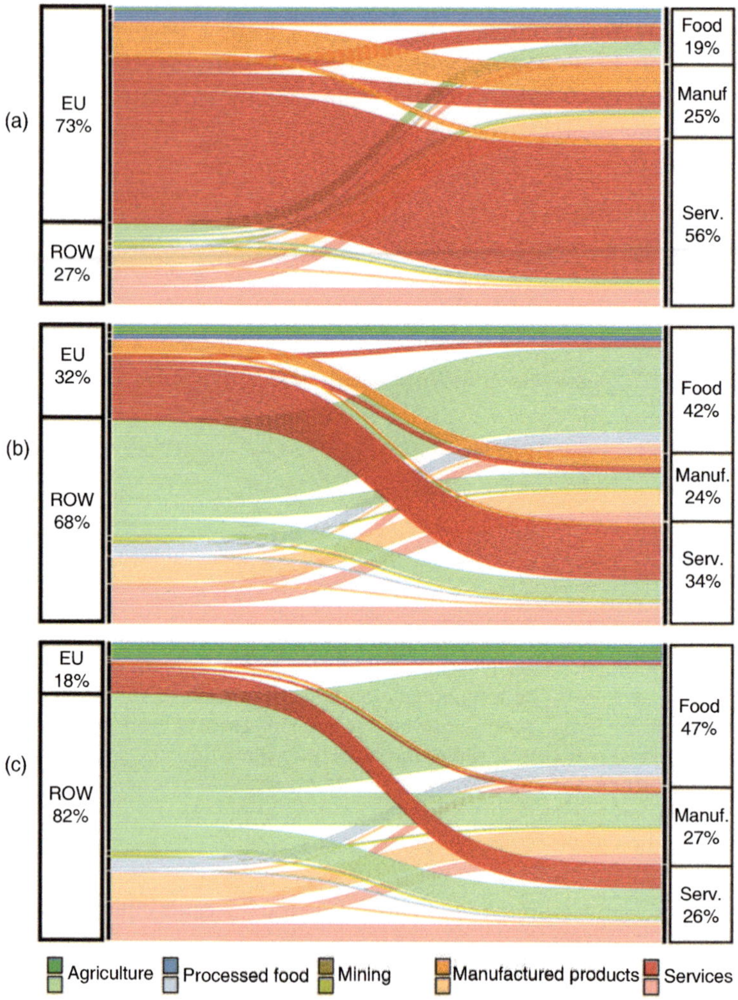

Figure 8.4 Flows of embodied wages (a), total employment (b) and vulnerable employment (c) associated with household consumption in the EU in 2011. On the left, and identified by colours, are the origin of the flows, where the workers were employed, aggregated in five industries (agriculture, processed food, mining, manufactured products and services) and two regions (EU and the rest of the world). On the right are the products consumed in the EU, aggregated in three categories (food, manufactured products and services).

8.4.1 Socioeconomic Footprints of Food, Clothes and Electronic Products

The distribution of wages, total employment and vulnerable employment presents some differences from those related to total consumption. Figure 8.5 shows the footprints for wages (left), total employment (centre) and vulnerable employment (right) for food (a), clothes (b) and electronic products (c) consumed by EU citizens. Overall, the difference between wages in the EU and other regions decreased, and by 2007, about half of the wages associated with the production of these goods were paid to workers outside the EU. For workers in total and in vulnerable employment, on the other hand, the pattern is very different. While total and vulnerable employment created in the EU presents a slight decrease in the period, the majority of the work—and especially of vulnerable work—remains outside the region. It is interesting to note how the production of clothes and electronics became increasingly outsourced after 2000, noticeable by a decrease in EU wages at the same time that the footprint for wages and workers outside the EU increased more intensively, widening the gap between the share of the labour footprint that originates inside and outside the EU. For food production, most of this outsourcing appears to have taken place before 1995.

The financial crisis of 2007–2008 affected all footprints because of the reduction of consumption by EU households. It can be noted that while the economic crisis had an effect on wages in the EU, the reduction of footprints for both wages and employment was more accentuated outside the region. The reduction of European consumption, thus, affected more workers outside the EU than those within the region.

But how much of total spending and total socioeconomic footprints are associated with these products? In 2011 the consumption of food products occupied about 16% of the average EU household budget and accounted for about the same share in wages footprints but corresponded to over 40% of employment footprint and almost 50% of all workers in vulnerable conditions associated with EU consumption. The importance of food products to socioeconomic footprints is paramount. First, food is continuously consumed by all citizens, and unlike other products, the purchase of food per capita represents low elasticity compared to other products [11] and the purchase of new products cannot be substituted by the consumption of second-hand ones. Second, agriculture is responsible for one-third of global employment [12] and is considered a key sector for poverty reduction [13]. Clothes and electronics corresponded to smaller shares of expenditure (3.9% and 2.6%, respectively) and of wages and employment footprints, as can be seen in Table 8.1.

Figure 8.5 Footprints for food (a), clothing (b) and electronic products (c) consumed by European households for wages, in constant purchasing power parity (PPP, 2011 international dollars, left), total employment (centre) and vulnerable employment (right). Line plots represent the total annual footprint from 1995 to 2011 (continuous lines), part of the footprint that happened in the European countries (dashed lines) and part of the footprint outside the European Union (dotted lines).

Table 8.1 Footprints for European household consumption in 2011 for all products and the share of footprints due to the consumption of food, clothing and electronic products

Expenditure and Footprints for European Household Consumption				
	All products	**Food**	**Clothes**	**Electronics**
Household expenditure	5,796,410	907,873	226,817	147,894
(millions of euros)	100%	15.7%	3.9%	2.6%
Wages	3,894,433	725,144	204,079	121,342
(millions of 2011 international dollars)	100%	16.5%	5.2%	3.1%
Employment	223,788	94,942	13,507	5503
(1000 person-years)	100%	42.4%	6.0%	2.5%
Vulnerable employment	85,583	40,529	7306	2220
(1000 person-years)	100%	47.4%	5.2%	2.0%

8.4.2 What Are the Impacts of Buying New Food, Clothing and Electronic Products?

The impacts in the supply chain of food, clothes and electronic products are illustrated in Fig. 8.6. It shows the total wages, employment and vulnerable employment embodied in the supply chain of spending 1 million euros. By converting footprints of products to the same unit—common expenditure instead of total consumption—we can compare the potential to generate income and jobs and to employ people in vulnerable conditions around the globe.

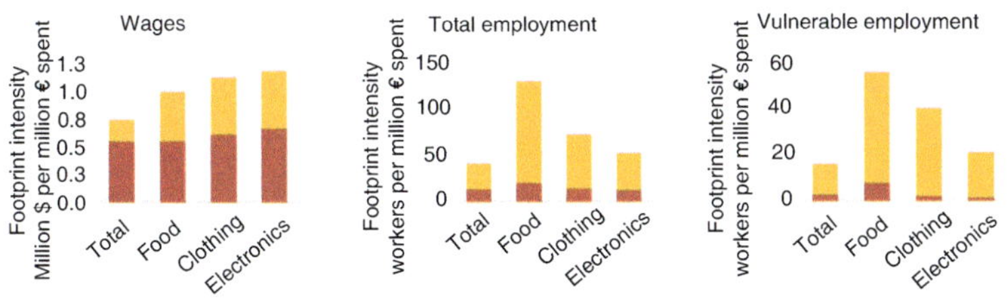

Figure 8.6 Footprint intensity in 2011, in impacts per million euros spent, for wages (left), total employment (centre) and vulnerable employment (right). The intensity of production within the European Union is represented by red bars, and the intensity of production outside the European Union is denoted by yellow bars.

Two things are evident in the figure. First, the purchase of food generates more jobs than other products, mostly outside the EU. Second, for every euro spent in products by EU consumers, 0.6–0.7 international dollars (or 0.4–0.5 euros) are paid to European workers. For food, clothing and electronic products, this value represents nearly half of the wages in the upstream supply chain of these products. However, European workers are paid, on average, 4–6.5 times more than foreign workers, adjusted to PPP. That means that most of the employment created in other regions is in low-paid work, mostly in vulnerable conditions. Workers in vulnerable conditions outside the EU represent 44%, 63% and 49% of all employment in the supply chains of food, clothing and electronic products, respectively. Workers in the EU, on the other hand, tend to have formal employment agreements and only 16% of workers in the supply chain of clothes and 13% of those in the supply chain of electronic products are in vulnerable conditions.

8.5 Conclusion

The economy has become more and more globalised in the last decades. Progress on transport and communication technologies means that a new labour organisation has unfolded in the global economic network. This means that production has become less centralised and supply chains have become more complex. Increasing specialisation and outsourcing means that our consumption creates social and environmental impacts worldwide, be it in the form of environmental and social damage in countries with reduced regulation or in the form of poverty alleviation and inclusion of new entrepreneurs in global markets.

The consumption of goods in Europe creates millions of jobs within and outside the EU, most of them in the supply chain of food products. Wages are concentrated in the EU, with over three-quarters of wages paid to workers who live within the EU, while 68% of the total employment and 82% of all workers in vulnerable conditions associated with EU consumption were outside the region. Food products are an example that occupied about 16% of the average EU household budget and accounted for about the same share in wage footprints but corresponded to over 40% of the employment footprint and almost 50% of all workers in a vulnerable condition associated with EU consumption. We have seen that over time, there has been strong growth in outsourcing, particularly recently in clothes and electronics, noticeable by a decrease in the EU wage footprint at the same time as the footprint for wages and workers outside the EU increased more intensively. For food production, most of this outsourcing appears to have taken place before 1995. The shock of the financial crisis, which saw a much bigger drop in employment impacts outside the EU, also shows the dependency problem that is created for poorer regions related to trade.

As half of all workers involved in supply chains of products consumed in the EU are in vulnerable conditions, we need to consider the impact of our consumption on the livelihoods of these people. While the presence of these workers—mostly own-account and family workers—in global supply chains is a step towards the inclusion of people in the job market, the increase of vulnerable workers relative to overall employment within the employment footprint of the EU is of concern. While the wage gap between workers within and outside the EU has decreased slightly, the sheer volume of employment and undesirable forms of labour conditions mean that we need to be actively engaged with, in particular, the agricultural labour force in other regions.

References

1. Tukker, A., Bulavskaya, T., Giljum, S., de Koning, A., Lutter, S., Simas, M., Stadler, K. and Wood, R. (2014). *The Global Resource Footprint of Nations: Carbon, Water, Land and Materials Embodied in Trade and Final Consumption Calculated with EXIOBASE 2.1* (Netherlands Organisation for Applied Scientific Research; Leiden University; Vienna University of Economics and Business; Norwegian University of Science and Technology: Leiden, the Netherlands; Delft/Vienna, Austria; Trondheim, Norway), http://www.exiobase.eu/downloads/creea-booklet.
2. Simas, M., Wood, R. and Hertwich, E. (2015). Labor embodied in trade: the role of labor and energy productivity and implications for greenhouse gas emissions, *Journal of Industrial Ecology*, **19**(3), 343–356.
3. The World Bank. (2016). *World Development Indicators*, http://data.worldbank.org/.

4. Wiedmann, T. and Barrett, J. (2013). Policy-relevant applications of environmentally extended MRIO databases: experiences from the UK, *Economic Systems Research*, **25**(1), 143–156.
5. McBain, D. (2015). Is social footprinting relevant to industrial ecology?, *Journal of Industrial Ecology*, **19**(3), 340–342.
6. International Labour Organization. (2011). *Trade and Employment: From Myths to Facts* (International Labour Organization, Geneva), http://www.ilo.org/wcmsp5/groups/public/@ed_emp/documents/publication/wcms_162297.pdf.
7. Dollar, D. and Kraay, A. (2004). Trade, growth, and poverty, *Economic Journal*, **114**(493), F22–F49.
8. Alsamawi, A., Murray, J., Lenzen, M., Moran, D. and Kanemoto, K. (2014). The inequality footprints of nations: a novel approach to quantitative accounting of income inequality, *PLOS ONE*, **9**(10), e110881.
9. Simas, M., Golsteijn, L., Huijbregts, M., Wood, R. and Hertwich, E. (2014). The 'bad labor' footprint: quantifying the social impacts of globalization, *Sustainability*, **6**(11), 7514–7540.
10. Wood, R., Stadler, K., Bulavskaya, T., Lutter, S., Giljum, S., de Koning, A., Kuenen, J., Schütz, H., Acosta-Fernández, J., Usubiaga, A., Simas, M., Ivanova, O., Weinzettel, J., Schmidt, J. H., Merciai, S. and Tukker, A. (2015). Global sustainability accounting: developing EXIOBASE for multi-regional footprint analysis, *Sustainability*, **7**(1), 138–163.
11. Hertwich, E. G. and Peters, G. P. (2009). Carbon footprint of nations: a global, trade-linked analysis, *Environmental Science & Technology*, **43**(16), 6414–6420.
12. International Labour Organization. (2009). *Global Employment Trends: January 2009* (International Labour Organization, Geneva).
13. Cervantes-Godoy, D. and Dewbre, J. (2010). *Economic Importance of Agriculture for Poverty Reduction*, Working paper no. 23 (OECD, France), https://www.oecd.org/tad/44804637.pdf.

Chapter 9

Assessing the Structure and Social Effects of China's Provincial Labour Landscape

Yafei Wang and Lixiao Xu
School of Statistics and Institute of National Accounts, Beijing Normal University, Beijing 100875, China
ywang@bnu.edu.cn

9.1 Introduction

We are presenting an in-depth study of the provincial labour landscape of China. For the first time, the provincial particulars of employment and income, as well as the effects that economic interdependencies and trade links have on these, are analysed.

We are applying three different measures to the workforce—inequality according to the Gini index, average wages and number of employed individuals—and focus on each measure from a consumer's and a producer's perspective.

Income inequality is one of the most important issues to be addressed by countries in order to improve their people's living

The Social Effects of Global Trade
Edited by Joy Murray, Arunima Malik, and Arne Geschke

ISBN 978-981-4774-55-0 (Hardcover), 978-020-3728-71-0 (eBook)
www.panstanford.com

quality and make societies more sustainable. Concern has been growing about income inequality as high levels of inequality generate high costs for society, dampen social mobility, undermine the labour market prospects of vulnerable social groups and create social unrest [1]. *Income Inequality: The Gap between Rich and Poor*, published by the Organisation for Economic Cooperation and Development (OECD), shows the growing income gap between the richest and the poorest in OECD countries. Income inequality has been rising not only in many developing countries but also in many developed ones. OECD data show that in China, which is the largest developing country, income inequality has risen by 1.6% on average per year over the last two decades [1].

The Gini index, expressed as a Gini ratio or a normalised Gini coefficient, is the most commonly used measure of inequality. It was developed by the Italian statistician and sociologist Corrado Gini and publicised in his paper "Variability and Mutability" in 1912 [2]. It measures the statistical dispersion intended to represent the income distribution of a given country's population. A coefficient of 0 represents perfect equality; a coefficient of 1 represents total inequality, where one person has everything and the rest of the population has nothing.

The footprint measure has been developed in recent years as an appropriate tool for tracing the whole supply chain of goods and services along the production system driven by different categories of final demand. It can be calculated for various satellite indicators such as carbon dioxide emission, water, material flows and biodiversity using a consumption-based accounting framework based on input-output (IO) analysis [3–7]. The inequality footprint, a new concept proposed by Alsamawi et al., is used to calculate the Gini index of footprinting results, that is, the income inequality of the workforce directly and indirectly required to satisfy the consumption of a given population [8, 9].

China is currently the world's second-largest economy, behind the United States, and its development depends strongly on its provincial economic growth and domestic inter-provincial trade linkages. However, there exists significantly unbalanced development across Chinese provinces because of policy preferences, industrial foundations and natural resource distribution [10]. This leads to considerable disparities in Chinese provinces. This chapter aims to measure the income inequality footprint of China's provinces for the year 2010 using a Chinese multi-regional input-output (MRIO) table. It provides an inequality analysis from a consumption-based perspective.

9.2 Methodology and Data

9.2.1 Methodology

The Gini index and IO analysis combine in the methodology to estimate the income inequality between provinces at the sectoral level. The Gini index is an essential concept to estimate the income inequality, which traditionally depicts the relationship between cumulative income and corresponding cumulative population. IO analysis was developed by the Nobel Prize winner Wassily Leontief to analyse the interdependence of sectors in a given economy [11]. With MRIO tables, the interdependence among sectors within one region can extend to include inter-regional linkages.

The inequality footprint applies the concept of the Gini index to the income distribution of an IO-based employment footprint. We evaluate the inequality footprint for each Chinese province according to the following steps:

1. **Calculate the provincial Gini indices for the year 2010:** There are no official Gini indices at the provincial level published by central or local governments in China. Thus we calculate the Gini index for every Chinese province according to the study by Tian [12].
2. **Calculate the regional employment distributions by income classes using the Gini index and income distribution:** We split the total number of employees into five income classes (as defined by the Chinese Urban Household Survey [13]) using the Gini index.
3. **Reconcile the sector classifications between the employment and income data and the Chinese MRIO table:** We align the employment sectors with MRIO by using a prorating approach based on an in-house binary concordance matrix, which was prorated using the sectoral output of each province as a proxy.

4. **Calculate employment and income footprints for each province:** We use the Leontief demand-driven model [14, 15], where f indicates the social footprint with either the employment or the income footprint, q represents the intensity of employment used or income generated from producing one unit of total output and L is the Leontief inverse matrix and is the diagonal vector of the final demand y.
5. **Calculate the Gini index for the footprints:** Now that the footprints for the different income classes are available, the Gini index for this set of footprints can be calculated. This Gini index for the footprint is what we refer to as the *inequality footprint.*

It is important to point out that due to its construction, the inequality footprint can be directly compared to any Gini index.

Despite the fact that the Gini index and the inequality footprint are comparable quantities, it is important to point out the differences. The Gini index is commonly used as a measure for the inequality within the entire workforce of a specific country. The inequality footprint—despite measuring the inequality within a group of workers, like the Gini index does—applies this measure to all the workers who are associated with the production of a particular final demand regardless of where they are located.

9.2.2 Data

The Chinese MRIO table features 30 provinces, with 30 sectors per province, and was taken from the work of Wang et al. [16] (see Table A.1 in the Appendix). The employment data for each province was separated into five income classes using data from *The 2011 Provincial Statistical Yearbooks* [17]. Following this data preparation, the provincial Gini indices were calculated. The total employment and total income at the sectoral level for each province are both obtained from *The 2011 Population and Employment Statistics Yearbook*, published by the National Bureau of Statistics of China [18]. The units of the employment and income are full-time equivalent (FTE) and 1000 Chinese yuan (CNY), respectively.

9.3 Results

9.3.1 Inequality Footprint of Provinces

Income inequality exists both between countries and domestically between regions of a particular country. Income inequality between countries is attributed to conflict, governance and possession of natural resources, technological change, jobs, transportation costs and globalisation [9]. At a domestic level, income inequality between regions within one country is because of geographical locations, policy preferences, industrial foundations and natural resource endowment [10]. A global income inequality footprint shows the link that each country's demand has with the income distribution within the workforce elsewhere in the world that is required to meet this demand. Similarly, a domestic regional inequality footprint links each region's demand to the associated income distribution in the workforce associated to this demand in the rest of the country. Income distribution from other countries or regions is the result of international or inter-regional trade. Therefore resulting income differences between countries or regions within one country are the source of the income inequality footprint.

Figure 9.1 presents the inequality footprint and Gini indices of 30 Chinese provinces. The figure shows the inequality footprint for each province above the midline and the domestic Gini indices for each province below the midline. The domestic Gini indices were directly derived from employment and income data. The key difference is that the footprint encompasses all employment that is associated with the particular province's consumption, whether this employment occurs domestically or in other provinces, whereas the domestic Gini index only considers the domestic workforce. Thus the difference between the concepts of an inequality footprint and the domestic Gini index is the same as the difference between the income generated by employment embodied in goods and services imported from other provinces and the income generated by domestic employment. In 2010, the inequality footprints of 22 out of 30 provinces were higher than their domestic Gini indices. This means that for the majority of provinces, the domestic workforce experiences more equal working conditions than the workforce captured by the inequality footprint. For example, Hebei shows the greatest difference between the inequality footprint (0.2682) and domestic inequality (0.2174) being 0.0508. This could be interpreted as Hebei outsourcing its inequality, given the workforce associated with Hebei's footprint experiences less equality than Hebei's domestic workforce. But the inequality footprint does not provide any information on the reasons for this observation, and neither is inequality a necessary input into a given product, compared, for example, to energy, which produces emissions that are encompassed by a carbon footprint. Hence the interpretation of the inequality footprint differs from that of more conventional footprints (such as emissions, water and land). Where conventional footprints indicate a causal relationship between final demand and the footprint (the final demand causes emissions and land and water use along the supply chains), the inequality footprint does not imply the same causal effect. Hence, for inequality footprints, further analysis is required to identify whether causal effects between Hebei's demand and an increased level of inequality along the supply chains compared to the domestic level of inequality exist. Similar observations are made for central and west China, such as Jiangxi, Anhui, Shanxi, Inner Mongolia, Xinjiang and Qinghai. The inequality footprints of these provinces are higher than their domestic Gini indices.

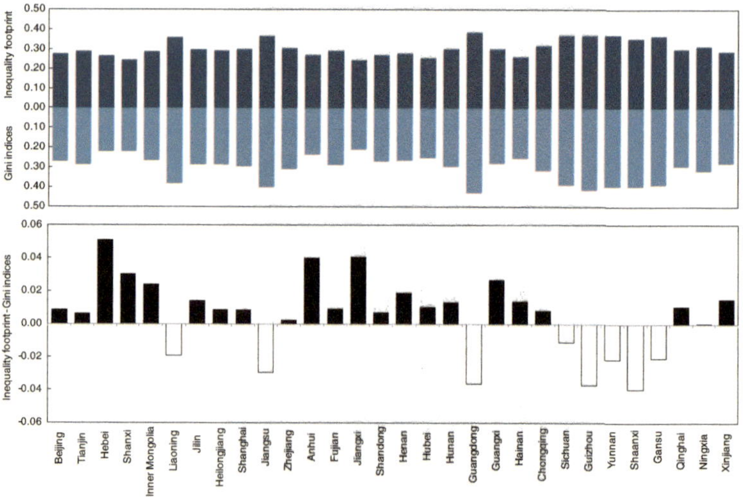

Figure 9.1 Inequality footprints and Gini indices of 30 Chinese provinces for 2010. (Top) Gini indices indicating domestic inequality and inequality footprints, including the domestic and inequality footprints embodied in the import of goods for each province. (Bottom) Differences between inequality footprints and Gini indices, that is, inequality embodied in the import of goods for each province.

9.3.2 Average Wages and Wage Footprints of Sectors

To shed more light on the observations on provincial inequality footprints presented in the last section, we will now focus on the wage footprints of the different sectors for each region. We define the average wage as the total income divided by the total employment FTE. This concept can be applied to the domestic workforce as well as to the employment and income footprint, thus providing us with another metric to compare the domestic- and footprint-related workforce. Figure 9.2, left panel, shows sectoral average wages within each province. From the sectoral rows, it is clear that the average wages of Extraction of Petroleum and Natural Gas, Fuel and Chemicals, Electricity and Heat Power, Post and Telecommunications, and Financial Intermediation and Business Activities in most provinces are higher than those of other sectors. Cells with very high wages are in deep blue, such as Extraction of Petroleum and Natural Gas, Electricity and Heat Power, Post and Telecommunications, Financial Intermediation and Business Activities in Shanghai, Food and Tobacco in Zhejiang, Fuel and Chemicals in Fujian and Post and Telecommunications in Beijing (Fig. 9.2, left panel).

The right panel of Fig. 9.2 shows the average wage of the sectoral footprint for each province (referred to as the sectoral wage footprints). In comparison to sectoral average wages, sectoral wage footprints show fewer overall differences for almost every sector. For example, Qinghai features both very high and very low domestic average wages across its sectors, whereas the sectoral wage footprints are much more homogenous. This indicates that the sectoral employment footprints are generally evenly spread across the different income classes. Hence, both high-income earners and low-income earners are part of the footprint, leading to a more homogenous distribution of sectoral wage footprints. This, in return, indicates that the inequality footprints that were presented in the previous section refer to employment footprints that span across the entire income range and approximately similar ratios (similar percentages of low-income earners and high-income earners are present in all sectoral footprints regardless of the average wage of the sector domestically).

There are some exceptions from these observations. Figure 9.2 also shows that some sectors with high domestic wages can have a relatively low wage footprint. See, for example, the Extraction of Petroleum and Natural Gas in Shanghai and Electricity and Heat Power in Beijing, Tianjin and Shanghai. This implies that there are, in fact, sectors that rely on inputs that are associated with lower wages.

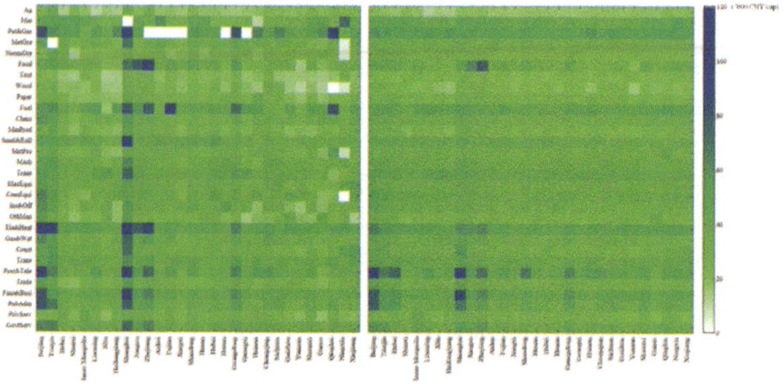

Figure 9.2 Sectoral average wages (left) and average wage footprints (right) of 30 provinces. Average wage equals income/employment in 1000 CNY per capita. The row axis shows the sector classification (the sectoral names are listed in Table A.1), and the vertical one represents the provinces. Ten sectoral values are zeros (white colour) (left). These sectors are as follows: Mining and Processing of Metal Ores of Tianjin; Mining and Washing of Coal of Shanghai; Extraction of Petroleum and Natural Gas of Zhejiang, Anhui, Fujian, Jiangxi, Hunan and Guangxi; Wood of Qinghai; and communication equipment and computers and other electronic equipment of Ningxia.

Some provinces do not produce certain goods, which corresponds to no income of wages in this sector domestically (cells with the white colour in Fig. 9.2). For example, there is no Extraction of Petroleum and Natural Gas in Zhejiang, Anhui, Fujian, Jiangxi, Hunan and Guangxi. These provinces import these kinds of goods from other provinces, resulting in average wage footprints from these sectors that are not matched by similar domestic wages in those sectors.

9.3.3 Supply Chains of Embodied Employment in Guangdong

We take Guangdong as an example to assess how employment is embodied in inter-provincial commodity flows and thus how the receiving province of associated commodity flow relies on employment in the exporting province. Guangdong, a typical manufacturing hub in China, has both the highest domestic Gini index (0.4227), making it the least equal province, and the highest inequality footprint (0.3884). We extract supply chains that are linked to Guangdong's final consumption, identify the provinces and sectors within these provinces in which the employment related to Guangdong's final consumption occurs and show which economic sectors within Guangdong receive the final product of each supply chain (Table 9.1).

The highest sectoral employment footprints that occur outside Guangdong are associated with similar sectors across numerous provinces:

- Mining and Washing of Coal (e.g., Shanxi, Henan, Hebei, Shandong and Guizhou), Government Services (e.g., Shanxi, Henan and Guizhou), Smelting and Rolling (e.g., Yunnan) and Electricity and Heat Power (e.g., Yunnan) supplying to Construction activities
- Electrical Machinery and Equipment (e.g., Zhejiang); Government Services (e.g., Yunnan); Smelting and Rolling (e.g., Yunnan); Mining and Washing of Coal (e.g., Shanxi); and Communication Equipment, Computers and Other Electronic Equipment (e.g., Jiangsu) supplying to Electrical Machinery and Equipment in Guangdong

The top-three supply chains are Mining and Washing of Coal in Shanxi supplying to Construction in Guangdong; Electrical Machinery and Equipment in Zhejiang supplying to Electrical Machinery and Equipment in Guangdong; and Communication Equipment, Computers and Other Electronic Equipment in Jiangsu supplying to Communication Equipment, Computers and Other Electronic Equipment in Guangdong. Respectively, these provide 170.93 thousand FTE (1.87%), 84.00 thousand FTE (0.92%), and

80.86 thousand FTE (0.88%) of the total employment footprint in Guangdong (Table 9.1).

Table 9.1 Top 20 sectors sorted by embodied employment of Guangdong

Rank	Production sectors	Consumption sectors	Embodied employment ('000 FTE)
1	Shanxi: Mining and Washing of Coal	Construction	170.93
2	Zhejiang: Electrical Machinery and Equipment	Electrical Machinery and Equipment	84.00
3	Jiangsu: Communication Equipment, Computers and Other Electronic Equipment	Communication Equipment, Computers and Other Electronic Equipment	80.86
4	Henan: Mining and Washing of Coal	Construction	74.92
5	Shanxi: Government Services	Construction	65.93
6	Hebei: Mining and Washing of Coal	Construction	59.69
7	Yunnan: Smelting and Rolling	Construction	48.29
8	Yunnan: Smelting and Rolling	Electrical Machinery and Equipment	44.85
9	Shanxi: Mining and Washing of Coal	Electrical Machinery and Equipment	38.94
10	Henan: Government Services	Construction	38.47
11	Yunnan: Government Services	Electrical Machinery and Equipment	38.05
12	Guizhou: Government Services	Construction	37.35
13	Shandong: Mining and Washing of Coal	Construction	37.08

Rank	Production sectors	Consumption sectors	Embodied employment ('000 FTE)
14	Guizhou: Mining and Washing of Coal	Construction	37.08
15	Jiangsu: Communication Equipment, Computers and Other Electronic Equipment	Electrical Machinery and Equipment	35.57
16	Yunnan: Electricity and Heat Power	Construction	34.80
17	Chongqing: Transport Equipment	Transport Equipment	34.22
18	Fujian: Communication Equipment, Computers and Other Electronic Equipment	Communication Equipment, Computers and Other Electronic Equipment	34.20
19	Hunan: Communication Equipment, Computers and Other Electronic Equipment	Communication Equipment, Computers and Other Electronic Equipment	33.28
20	Fujian: Textiles and Wearing Apparel	Textiles and Wearing Apparel	31.10

The above sectors mainly provide energy, raw materials and services for Guangdong's manufacturing and construction production. Lower wages in these sectors flowing from other provinces will result in a relatively low-income footprint compared to the domestic income in Guangdong's corresponding sectors.

9.4 Conclusion

In this chapter, we carried out an analysis of what affects the economic interdependencies and trade links between the 30 Chinese provinces have on the employment and income structure. We applied three different measures: (i) assessment of inequality across the workforce, (ii) assessment of the average wages for each sector in each province and (iii) the number of FTE to assess the employment for each sector and each province.

We applied all three measures for the domestic workforce to the workforce associated with the consumption footprint of each province. This is the first time that an assessment of China's regional employment and income distributions has been carried out from the consumption-based perspective.

Our study in this chapter provided us with a new perspective from which to understand the provincial labour landscape in China. Regional inequality can be observed not only from the domestic Gini index but also from the consumption perspective. More attention should be paid to understanding how the increasing inter-provincial trade fosters inequality and income disparity across the different provinces. The data we presented in this chapter will be able to assist further studies with this focus in the future.

Additionally, inequality footprints can be used as an indicator to monitor and analyse the current situation of and the trend in income disparity and help in understanding domestic provincial inequality by tracking embodied employment and wage footprints through the inter-provincial products' trade.

Appendix

Table A.1 List of 30 sectors in this chapter

No.	Sector names	Abbreviation
1	Agriculture	Ag
2	Mining and Washing of Coal	Min
3	Extraction of Petroleum and Natural Gas	Pet&Gas
4	Mining and Processing of Metal Ores	MetOre
5	Mining and Processing of Nonmetal Ores and Other Ores	NonmOre
6	Food and Tobacco	Food
7	Textiles and Wearing Apparel	Text
8	Wood	Wood
9	Paper	Paper

No.	Sector names	Abbreviation
10	Fuel and Chemicals	Fuel
11	Chemicals	Chem
12	Nonmetallic Mineral Products	MinProd
13	Smelting and Rolling	Smel&Roll
14	Metal Products	MetPro
15	Machinery Equipment	Mach
16	Transport Equipment	Trans
17	Electrical Machinery and Equipment	ElecEqui
18	Communication Equipment, Computers and Other Electronic Equipment	ComEqui
19	Measuring Instruments and Machinery for Culture Activity and Office Work	Ins&Off
20	Other Manufacture	OthMan
21	Electricity and Heat Power	Ele&Heat
22	Gas and Water	Gas&Wat
23	Construction	Const
24	Transport and Storage	Trans
25	Post and Telecommunications	Post&Tele
26	Wholesale and Retail Trades, Hotels and Restaurants	Trade
27	Financial Intermediation and Business Activities	Fina&Busi
28	Public Administration	PubAdm
29	Private Services	PrivServ
30	Government Services	GovtServ

References

1. Keeley, B. (2015). *Income Inequality: The Gap between Rich and Poor* (OECD, France).
2. Gini, C. (1912). Variability and mutability, contribution to the study of statistical distributions and relations, *Studi Economico-Giuricici della R. Universita de Cagliari.*

3. Feng, K., Davisb, S. J., Sun, L., Li, X., Guan, D., Liu, W., Liu, Z. and Hubacek, K. (2013). Outsourcing CO_2 within China, *Proceedings of the National Academy of Sciences of the United States of America*, **110**, 11654–11659.
4. Feng, K., Hubacek, K., Pfister, S., Yu, Y. and Sun, L. (2014). Virtual scarce water in China, *Environmental Science & Technology*, **48**, 7704–7713.
5. Lenzen, M., Moran, D., Bhaduri, A., Kanemoto, K., Bekchanov, M., Geschke, A. and Foran, B. (2013). International trade of scarce water, *Ecological Economics*, **94**, 78–85.
6. Lenzen, M., Moran, D., Kanemoto, K., Foran, B., Lobefaro, L. and Geschke, A. (2012). International trade drives biodiversity threats in developing nations, *Nature*, **486**, 109–112.
7. Wiedmann, T. O., Schandl, H., Lenzen, M., Moran, D., Suh, S., West, J. and Kanemoto, K. (2015). The material footprint of nations, *Proceedings of the National Academy of Sciences of the United States of America*, **112**, 6217–6276.
8. Alsamawi, A., Murray, J. and Lenzen, M. (2014). The employment footprints of nations: uncovering master-servant relationships, *Journal of Industrial Ecology*, **18**(1), 59–70.
9. Alsamawi, A., Murray, J., Lenzen, M., Moran, D. and Kanemoto, K. (2013). The inequality footprints of nations: a novel approach to quantitative accounting of income inequality, *PLOS ONE*, **9**(10), e110881.
10. Guo, S. (2007). The internal and external imbalances of China's economy, *Economic Research Journal*, (12), 4–10 (in Chinese).
11. Miller, R. E. and Blair, P. D. (2009). *Input-Output Analysis: Foundations and Extensions* (Cambridge University Press, Cambridge).
12. Tian, W. (2012). The measurement and trend analysis of provincial income Gini index, *Economic Science*, (02), 48–59 (in Chinese).
13. National Bureau of Statistics of China. (2011a). *Chinese Urban Household Survey* (China Statistics Press, Beijing).
14. Leontief, W. (1949). Recent developments in the study of interindustrial relationships, *American Economic Review*, **39**(3), 211–225.
15. Leontief, W. and Strout, A. (1963). *Multiregional Input-Output Analysis* (Palgrave Macmillan, London).
16. Wang, Y., Geschke, A. and Lenzen, M. (2015). Constructing a time series of nested multiregion input-output tables, *International Regional Science Review*, doi:10.1177/0160017615603596.
17. National Bureau of Statistics of China. (2011c). *Provincial Statistical Yearbooks* (China Statistics Press, Beijing).
18. National Bureau of Statistics of China. (2011b). *Population and Employment Statistics Yearbook* (China Statistics Press, Beijing).

Chapter 10

Fairness and Globalisation in the Western European Clothing Supply Chain

Simon Mair, Angela Druckman and Tim Jackson
Centre for Environment and Sustainability, University of Surrey, Guildford, Surrey, United Kingdom
s.mair@surrey.ac.uk

In this chapter we use global multi-regional input-output analysis to explore how globalisation has impacted fairness along Western European clothing supply chains. Our analysis shows that while globalisation has made the Western European clothing supply chain 'fairer' by increasing employment opportunities and income for workers in Brazil, Russia, India and China (BRIC), it has failed to make the supply chain fair. Despite large increases in the labour compensation received by BRIC workers in the Western European clothing supply chain, labour compensation is still insufficient to support a decent standard of living and cannot, therefore, be considered fair.

The Social Effects of Global Trade
Edited by Joy Murray, Arunima Malik, and Arne Geschke

ISBN 978-981-4774-55-0 (Hardcover), 978-020-3728-71-0 (eBook)
www.panstanford.com

10.1 Fairness, Globalisation and the Good Life

Fairness is central to many, if not all, visions of 'the good life' [1, 2], but it is unclear whether globalisation makes the world a fairer place. Consequently, examining the relationship between globalisation and fairness can help us to understand what role globalisation might play in efforts to attain the good life. Proponents of globalisation argue that it increases the opportunities and income of many of the world's poorest people. Indeed, in real terms the incomes of those in many developing countries have risen faster than incomes in developed countries over the last two decades [3]. Conversely, critics argue that globalisation facilitates the exploitation of workers by large transnational corporations. They point to examples such as the Rana Plaza garment factory, which collapsed because building regulations were not enforced by politicians who feared that enforcing them would cause international clothing brands to source their clothes from elsewhere, taking jobs with them [4]. In the collapse, 1100 workers were killed and 2500 injured.

In this chapter we use results from our global multi-regional input-output analyses to explore the notion of fairness in the context of the Western European clothing supply chain [5–7]. In particular we focus on how globalisation has affected the fairness of labour compensation in the Western European supply chain. Clothing supply chains make a useful exploration ground for such issues because they have historically been at the forefront of globalisation and economic development and remain the focus of much conflict over globalisation and labour rights.

The chapter is structured as follows. In the next section we describe how globalisation has reshaped clothing supply chains. Then we show how labour compensation has been distributed across different groups of workers in the Western European clothing supply chain and how this is linked to globalisation. Finally we focus on workers in the Brazil, Russia, India and China (BRIC) countries. Yes, they are better off because of globalisation, but does this constitute fairness?

10.2 The Changing Geography of the Clothing Supply Chain

As discussed elsewhere in this book, globalisation post-1990 has been characterised by increasingly fragmented production. That is, since 1990 the number of countries involved in any given supply chain has tended to increase [8]. This is thought to have been driven by the vertical specialisation of production, a process in which firms in different countries specialise in discrete stages of the production process, leading to an increase in the international trade of intermediate goods [9, 10].

However, this general trend is not quite the whole picture—global trade structures are constantly shifting, shaped by changes in broader macroeconomic forces. From 1970, fragmentation of the clothing supply chain was exaggerated by trade policies. Most notably the Multi-Fiber Arrangement (MFA) regulated most global clothing trade until 2005 [11]. The MFA restricted access to developed country markets for developing country textile and clothing producers. This was achieved through negotiation of quotas, which limited imports of specific goods into developed countries from developing countries. The quota system caused producers to switch production to countries with no or unfulfilled quotas, exacerbating fragmentation [12]. In some cases this process created clothing industries in small developing countries almost overnight [11]. More recently, however, clothing supply chains have begun to consolidate and regionalise [13]. The year 1995 marked the beginning of the phasing out of the MFA, and since then the relative share of production by China and other major suppliers has grown at the expense of many of the smaller regions who benefitted from the MFA quota system [14].

These competing trends can be seen in Fig. 10.1 (based on the input-output analysis in Mair et al. [5]), which shows how the hours worked in the Western European clothing supply chain were distributed across three world regions between 1995 and 2009. Although the supply chain is undoubtedly global, the BRIC countries dominate the entire time series. Particularly striking is the strong rise in the amount of BRIC hours worked in the supply chain in the later stages of the MFA phase-out. Also clear is the steady fall in the share of hours worked in Western Europe as these countries moved to service-based economies. It is also worth pointing out the decrease in hours worked at the end of the time series. A substantial portion of this is likely to be due to reduced consumption in Western Europe following the 2008 recession, but some is also attributable to increased efficiencies to keep up with the demands of fast fashion, and some may be due to increased use of workers in countries (e.g., Bangladesh), for which we do not have data [6].

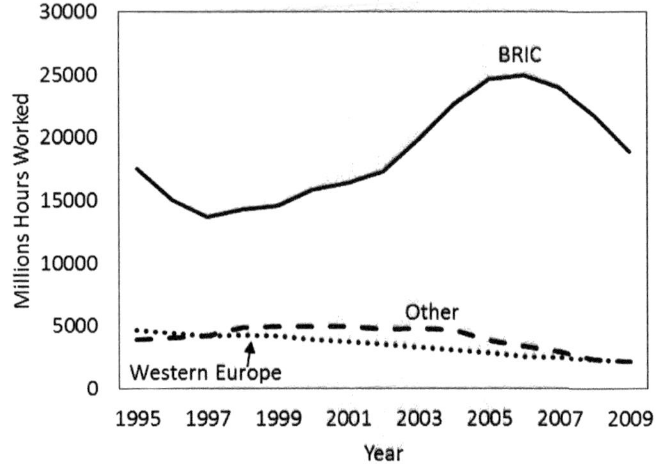

Figure 10.1 Hours worked in BRIC, Western Europe and other countries in the Western European clothing supply chain. Adapted from Ref. [5], Copyright (2016), with permission from Elsevier.

10.3 Has Globalisation Made the Western European Clothing Supply Chain Fairer?

What do these trends tell us about fairness? On the one hand, Fig. 10.1 appears to support the claim that globalisation has increased employment opportunities for poor workers. Western Europe has offshored many of its jobs, creating jobs elsewhere. In the clothing supply chain, Fig. 10.1 would suggest that BRIC workers have been the principal recipients. This is clearly beneficial to those workers who now have jobs in BRIC but did not previously. However, at least some of these jobs come at the expense of workers in smaller non-BRIC countries [15]. Moreover, the BRIC countries typically have low wages and relatively poor labour rights [16].

Trends in labour compensation also only partially support the idea that globalisation has made the Western European clothing supply chain fairer. Figure 10.2 shows the labour compensation received by BRIC and Western European workers in the Western European clothing supply chain between 1995 and 2009. Western Europe labour compensation fell, while BRIC labour compensation rose, mirroring the trends in hours worked. It is likely that a large portion of the changes in labour compensation can be explained by the changes in hours worked. From this, it follows that by facilitating the offshoring of labour-intensive work by affluent countries, globalisation increased employment and, therefore, the income of workers in poorer countries. It is worth noting, however, that in Fig. 10.2 labour compensation remains much higher in the Western European portions of the supply chain despite the much lower hours worked (Fig. 10.1), a potential inequity. Indeed, changes in hours worked do not fully explain the changes in labour compensation, as we show below by looking at changes in the hourly labour compensation rate.

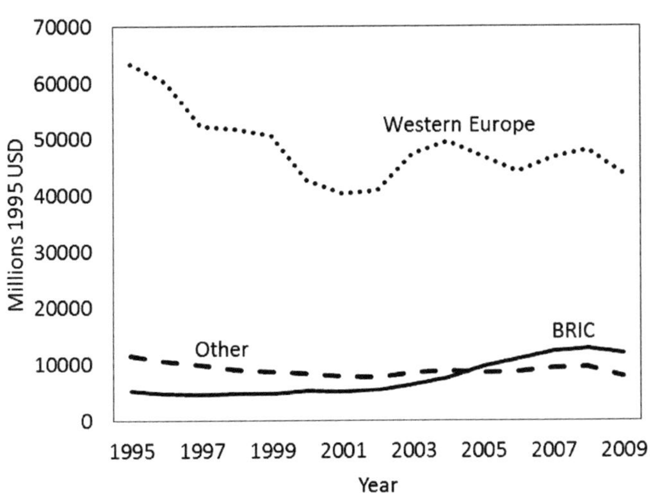

Figure 10.2 Labour Compensation in BRIC, Western Europe and other countries in the Western European clothing supply chain. Adapted from Ref. [5], Copyright (2016), with permission from Elsevier.

As with hours worked and labour compensation, looking at the changes in the hourly labour compensation rate of BRIC workers in the Western European clothing supply chain presents a mixed picture on fairness. It is clear that BRIC workers benefitted from globalisation: their hourly labour compensation doubled between 1995 and 2009 (from 1995 US$0.31 to US$0.62 per hour; Fig. 10.3). This is intuitive, because consolidation of the garment supply chain work in regional production centres has been associated with upgrading: the work is more skilled and commands higher wages [16]. Taken with the trends of increased hours and increased labour compensation, this suggests that the Western European clothing supply chain has become somewhat fairer, at least from the point of view of BRIC workers. However, perhaps the most striking element of Fig. 10.3 is the substantial difference between labour compensation rates in the BRIC and Western European parts of the Western European clothing supply chain. BRIC wages doubled between 1995 and 2009, but they remain around *30 times lower* than wages in Western Europe. This is likely a reflection of the fact that globalisation has seen affluent countries move low-skill, low-wage and insecure work to poorer countries, while largely retaining the higher-skilled and higher-wage work [7, 8]. Indeed, our model results show that most of the increase in BRIC working hours in the Western European clothing supply chain (Fig. 10.1) was either in the agricultural or textiles and clothing sectors, both of which are typically low-skill and low-wage sectors of the economy [5].

However, there are difficulties with comparing income across different countries, and these call into question the apparent unfairness of the large differences in labour compensation rates. It is commonly argued, for example, that although developing country wage rates appear low compared to those in affluent countries, this is because prices are much lower in developing countries, meaning that a low wage goes further. To address this argument we must estimate whether BRIC workers were paid enough to live a decent life, that is, whether they earned enough to meet their basic needs and fully partake in society [6, 7].

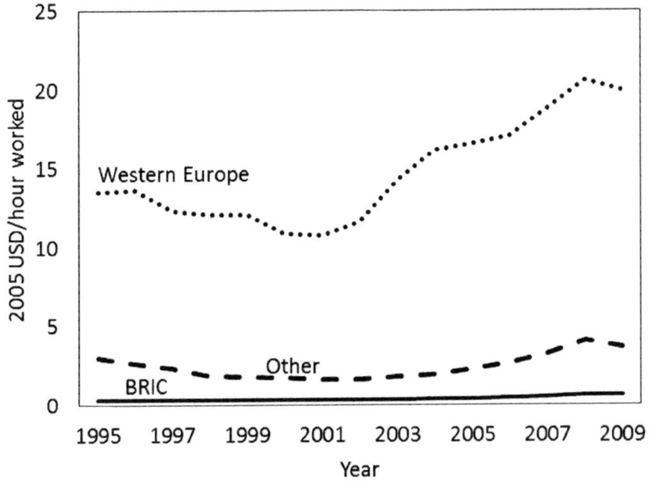

Figure 10.3 Labour compensation per hour in BRIC and Western Europe in the Western European clothing supply chain. Adapted from Ref. [5], Copyright (2016), with permission from Elsevier.

10.4 Does the Western European Clothing Supply Chain Support a Decent Life for BRIC Workers?

A living wage is an estimate of how much a worker would need to earn in order to have a basic but decent standard of living. Typically this means that workers are able to meet their basic material needs (shelter, adequate nutrition, etc.), support a small number of dependents and have a small discretionary income. The living wage concept is well established in both academic research and international law and is widely held to be a good benchmark of a fair wage [6, 7].

Here we use a living labour compensation (the living wage plus income taxes and social security payments) satellite account for the BRIC countries that is compatible with the World Input-Output Database. We use living labour compensation rather than living wages because it is directly comparable with labour compensation, which is a more conceptually robust way of measuring worker remuneration than wages [6]. Applying BRIC labour compensation in the Western European clothing supply chain allows us to estimate what labour compensation would have been *if BRIC workers could afford a decent standard of living*.

Figure 10.4 compares the actual and living labour compensation estimates for BRIC workers in the Western European clothing supply chain in 2005. Had BRIC workers received labour compensation that enabled them to afford a decent standard of living, the BRIC labour compensation bill in the Western European clothing supply chain would have approximately doubled. Put another way, in 2005 in the Western European clothing supply chain, the average *living* labour compensation rate for BRIC workers was approximately US$1 per hour compared with an *actual* labour compensation rate of approximately US$0.5 per hour. This suggests that BRIC workers in the Western European clothing supply chain (while being paid more fairly than in 1995) were still receiving an unfair wage—their wages could not support a reasonable standard of living.

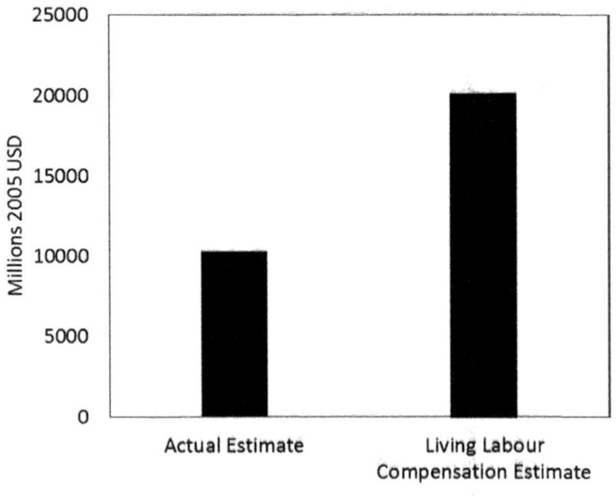

Figure 10.4 Change in the cost of labour in the BRIC parts of the Western European clothing supply chain associated with paying BRIC workers a living labour compensation rate.

Moreover, Fig. 10.4 provides a useful estimate of how much extra it would cost to make BRIC wages fair. In 2005, raising the labour compensation of BRIC workers in the Western European clothing supply chain to a level that would have supported a decent standard of living would have cost an additional US$10 billion. However, in 2005 Western European consumers spent around US$350 billion on clothing goods. All else being equal, paying BRIC workers in the Western European clothing supply chain a living labour compensation rate would have added approximately 3% to the average item of clothing purchased in Western Europe. Of course, all else is not equal: some taxes on the final product might increase in line with rising labour costs, and some goods are more exposed to BRIC labour than others (so the prices of these goods would increase by more than 3%). Nonetheless, this figure is in line with estimates from more detailed work [7, 17] and suggests that there is scope to pay living labour compensation in the Western European clothing supply chain. Furthermore, there is at least one case documented in the academic literature of one Western European company doing just this in its garment factories [18] and several major high-street brands have committed to investigating full-value chain living wage programs [19].

10.5 Towards Fair Globalisation?

The above analysis suggests that globalisation has made the Western European clothing supply chain fairer but not fair. Globalisation has created new jobs for workers in some poorer countries (notably the BRIC countries), but these came at the expense of jobs in other, smaller and arguably less economically developed, countries. Moreover, the jobs that were created in poorer countries were principally low-wage and low-skilled jobs. Consequently, despite substantial increases in labour compensation rates in the BRIC countries, workers still earn many times less than their Western European counterparts. Not only do they earn less, but our analysis suggests that the labour compensation received by BRIC workers in the Western European clothing supply chain was insufficient to support a decent standard of living and should, therefore, be considered unfair.

Consequently the analysis presented here highlights the need to explore ways to make things fairer. One such approach might be to increase labour compensation rates to a *living labour compensation* level that can support a decent life. Our analysis supports other work that suggests that this could be a feasible strategy as it would only carry a small relative price increase from the perspective of the final consumer, while greatly improving supply chain fairness and the quality of life of supply chain workers.

References

1. Perri, 6 and Christie, I. (1998). On the good life, in *The Good Life*, eds. Christie, I. and Nash, L. (Demos, London), pp. 3–20.
2. Jackson, T. (2016). *Prosperity without Growth: Foundations for the Economy of Tomorrow*, 2nd ed. (Routledge, London).
3. Milanovic, B. (2013). Global income inequality in numbers: in history and now, *Global Policy*, **4**(2), 198–208.
4. Taplin, I. (2014). Who is to blame? A re-examination of fast fashion after the 2013 factory disaster in Bangladesh, *Critical Perspectives on International Business*, **10**(1/2), 72–83.
5. Mair, S., Druckman, A. and Jackson, T. (2016). Global inequities and emissions in Western European textile and clothing consumption, *Journal of Cleaner Production*, **132**, 57–69.
6. Mair, S., Druckman, A. and Jackson, T. (In press). Investigating fairness in global supply chains: applying an extension of the living wage to the Western European clothing supply chain, *International Journal of Life Cycle Assessment*.
7. Mair, S. (2016). *Better Rather Than More? Exploring the Sustainability Implications of Paying a Living Wage in the Western European Clothing Supply Chain*, PhD thesis, University of Surrey.
8. Timmer, M. P., Erumban, A., Los, B., Stehrer, R. and de Vries, G. J. (2014). Slicing up global value chains, *Journal of Economic Perspectives*, **28**(2), 99–118.
9. Feenstra, R. and Hanson, G. (1996). *Globalization, Outsourcing, and Wage Inequality*, NBER working paper no. 5424, http://www.nber.org/papers/w5424 (accessed on 7 August 2014).
10. OECD. (2014). *OECD Factbook 2014: Economic, Environmental and Social Statistics* (OECD, France), p. 260.
11. OECD. (2004). *A New World Map in Textiles and Clothing Adjusting to Change: Adjusting to Change* (OECD, France), p. 234.
12. Dicken, P. (2011). *Global Shift: Mapping the Changing Contours of the World Economy*, 6th ed. (Sage), pp. 302–330.

13. Pickles, J., Plank, L., Staritz, C. and Glasmeier, A. (2015). Trade policy and regionalisms in global clothing production networks, *Cambridge Journal of Regions, Economy and Society*, **8**(3), 381–402.

14. Moazzem, K. G. and Sehrin, F. (2016). Economic upgrading in Bangladesh's apparel value chain during the post-MFA period: an exploratory analysis, *South Asia Economic Journal*, **17**(1), 1–21.

15. Pickles, J. (2012). Economic and social upgrading in apparel global value chains: public governance and trade policy, http://www.capturingthegains.org/pdf/ctg-wp-2012-13.pdf (accessed on 19 April 2016).

16. International Trade Union Confederation. (2015). ITUC global rights index: the world's worst countries for workers, http://www.ituc-csi.org/IMG/pdf/survey_global_rights_index_2015_en.pdf (accessed on 19 November 2015).

17. Miller, D. and Williams, P. (2009). What price a living wage? Implementation issues in the quest for decent wages in the global apparel sector, *Global Social Policy*, **9**(1), 99–125.

18. Egels-Zandén, N. (2015). The role of SMEs in global production networks a Swedish SME's payment of living wages at its Indian supplier, *Business & Society*, 1–38.

19. McMullen, A. and Majumder, S. (2016). *Do We Buy It? A Supply Chain Investigation into Living Wage Commitments from M&S and H&M: Labour behind the Label*, http://labourbehindthelabel.org/wp-content/uploads/2016/02/DoWeBuylt-spreadssml.pdf (accessed on 22 May 2016).

Chapter 11

Income Effects in Global Value Chains Driven by EU Exports

Ruslan Lukach and José M. Rueda-Cantuche[a]
European Commission, DG Joint Research Centre, Edificio Expo, Calle Inca Garcilaso, 3, E-41092 Seville, Spain
JoseM.RCantuche@ec.europa.eu

This chapter provides an overview of the main trends and patterns in income effects in global value chains driven by EU exports during the period 1995–2011. It makes extensive use of novel indicators illustrating the relationship between trade, employment and income (expressed as value added [VA]) for the EU as a whole and for each EU member state using the World Input-Output Database (WIOD) as the source for the data. The WIOD explicitly describes the interactions and interdependencies of the economic activities in different countries. The results shown in this chapter are focused on the income effects driven by the EU's exports to the rest of the world, with a geographical breakdown of the data that includes the 27 EU

[a]The views expressed herein are those of the authors and do not necessarily reflect an official position of the European Commission.

The Social Effects of Global Trade
Edited by Joy Murray, Arunima Malik, and Arne Geschke

ISBN 978-981-4774-55-0 (Hardcover), 978-020-3728-71-0 (eBook)
www.panstanford.com

member states (Croatia was not yet a member state in the period covered by this analysis, and the United Kingdom is still part of the EU), Australia, Brazil, Canada, China, India, Indonesia, Japan, Mexico, Russia, South Korea, Taiwan, the United States and an aggregate rest-of-the-world region.

These data allow us to examine for the first time the evolution of the VA (income) per EU member state embodied in all EU exports. Our time frame includes periods covering the global financial crisis and some years afterwards, shedding light on the interaction between trade and income flows. Furthermore, the current analysis offers new specific insights about the income that is generated elsewhere in the world by EU exports.

11.1 Introduction

Looking at the current international discussion regarding the future of the global economy, trade is one of the issues that naturally comes into sight as one of the main determining factors [1]. Over the past two decades, multinational enterprises have increasingly organised their production processes across the globe, in different countries. The automotive industry [2], the apparel industry [3] and the agri-food sector (e.g., Nutella®; see Ref. [4]) are good examples of this globalisation process. Following De Backer and Miroudot [4], a value chain identifies the full range of activities that firms undertake to bring a product or a service from its conception to its end use by final consumers. Many factors, such as technological progress, lower costs, access to resources and markets and trade policy reforms, have facilitated the geographical fragmentation of production processes across the globe. This international fragmentation of production is a powerful source of increased efficiency and firm competitiveness and has implications in many policy areas, starting with trade, investment and industrial development.

However, global value chains become increasingly dependent on complex production and investment networks that support commercial relations. This growing complexity makes it increasingly difficult to analyse and quantify the actual impacts that international trade has on the economic performance of national economies. This study provides better evidence to examine the position of countries within international production networks in terms of the income effects of the EU export activities within the EU and on the rest of the world.

The income flows in this study are represented by the flows in value added (VA), the difference between the value of output and the value of intermediate consumption of inputs. It can refer to a given industry or to an economy as a whole. It also equals the sum of labour compensation and return on capital (i.e., taxes less subsidies on production, consumption of fixed capital and net operating surplus, altogether), which is sometimes used as a proxy of the gross domestic product (GDP) when computed at the national level.

The GDP as a macroeconomic measure is not sufficient to provide an end measure of well-being but is rather to be considered as the means for achieving access to economic opportunities [5], which can be translated into improvements in the well-being of different population groups depending on the labour and capital income distribution patterns. Thus, the patterns of the VA flows driven by international trade allow us to obtain a clearer idea about the contribution of trade to forming the country's own GDP and the corresponding well-being.

The conceptual basis for the trade flows analysis based on international input-output relationships has been laid down in a number of papers focusing on empirical measurement of the VA of trade (e.g., see Refs. [6–10]).

Figure 11.1 shows various examples of the different transactions accounted for in both domestic (straight lines) and foreign (dotted lines) VA in EU exports. Germany may export domestically produced cars to China, thus generating VA in the EU (e.g., Germany). Alternatively, Germany may export engines to China, which in turn will produce Chinese cars to be exported to Japan or to be purchased by Chinese residents. Thus, in both cases, there would be VA generated in Germany and embodied in the Chinese cars sold in Japan or in China. It can be even more complicated. Germany may export vehicle parts to China, which will return to Germany in the form of vehicle engines. Ultimately, Germany would produce the car, which would be purchased by German residents. So, in the latter case, there would be two types of VA embodied in the German cars: that produced by manufacturers of vehicle parts and that related to the car manufacturing industry.

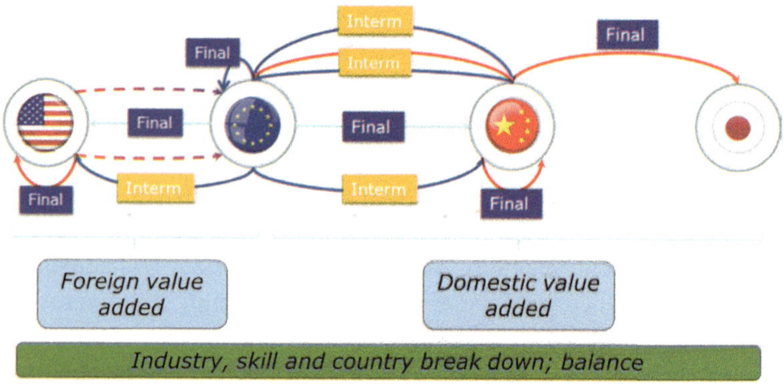

Figure 11.1 Example of value-added flows in final and intermediate products in global value chains as described by WIOD data. *Note*: The flow estimates can be calculated for different industries, labour skill levels and country breakdowns. The balance between domestic and foreign effects can also be computed.

On the other side, EU exports generate VA in other non-EU countries. Suppose Germany exports cars to the United States; it may happen that the engines have been, in turn, imported from the United States as well (or from elsewhere but for the sake of making the graph simpler, we have again chosen the same country). Thus, German exports of cars to non-EU countries would generate VA in other non-EU countries (dotted lines). It may also happen that the EU exports instead engines to the United States, which would produce cars to be sold domestically or abroad. In this case, the EU exports of engines may also generate VA in the United States through the imports of vehicle parts needed to produce engines.

Stehrer [11] developed two distinctive approaches to measuring the VA flows in international trade. The first approach uses the so-called *trade-in-VA approach*, which accounts for the VA of one country directly and indirectly contained in the final consumption of another country. The second, *VA-in-trade approach*, describes the VA contained in gross trade flows between two countries. This second definition allows us to consider such research questions as 'How much of VA from other countries is contained in the gross imports of one country?' and 'How much of foreign VA does the gross exports of a country embody?'

The data presented in this chapter follows the VA-in-trade approach and is based on several key concepts closely related to the literature on global value chains and the international fragmentation of production. To facilitate further discussion, we present the main relevant definitions and supply a typical illustration in Fig. 11.1. First, making an estimation of the VA generated by EU exports to the rest of the world in a given EU member state, we take into account that production is often organised along intra-EU value chains. Therefore, for each member state we calculate several value-chain effects: (i) a *domestic effect*, which captures the *national VA*, which is generated by its direct exports of goods and services to the rest of the world (excluding other EU countries) and (ii) a *spillover effect* or indirect effect, which captures the national VA that is engaged in the production of intermediate inputs to be used in other member states' exports to the rest of the world. Secondly, this analysis also accounts for the fact that EU production chains often extend well beyond EU borders. We introduce a concept of *foreign VA* to capture the VA created in upstream activities located outside the EU that feed inputs into the domestic production of EU exports.

This analysis uses the World Input-Output Database (WIOD) [12] and covers the period between 1995 and 2011 (although due to data constraints some data are only available up to 2009). This allows us to examine for the first time the evolution of the income flows generated by EU exports, including time after the outbreak of the global financial crisis. Furthermore, the current analysis offers more specific insights into some key EU bilateral trade relationships in terms of the income flows that they support.

The data presented in Ref. [13] feature a series of indicators to illustrate in detail the relationship between trade, employment and income for the EU as a whole and for each EU member state using the WIOD as the source for the data. All the indicators relate to the EU's exports to the rest of the world so as to reflect the scope of EU trade policy-making. Most indicators cover the period 1995–2011, but due to data constraints some are only available up to 2009. The geographical breakdown of the data includes the 27 EU member states (Croatia was not yet a member state in the period covered by this analysis), as well as Australia, Brazil, Canada, China, India, Indonesia, Japan, Mexico, Russia, South Korea, Taiwan, the United States and an aggregate rest-of-the-world region.

11.2 Domestic and Foreign Income Effects Generated by EU Exports to the Rest of the World

To obtain the EU and foreign income flows embodied in EU exports to the rest of the world, we have used a methodological approach based on a multi-regional input-output (MRIO) model together with the information provided by the WIOD. This type of analysis of the economic and environmental consequences of trade follows the example set by an established body of literature (for an extended overview see Refs. [14, 15]).

11.2.1 Value Added at Constant Prices (1995–2009)

According to the WIOD estimates, during the time period between 1995 and 2009, the VA in constant prices of 2009 (shown in Fig. 11.2) in the EU increased by EUR 2.4 trillion to a total of EUR 10.6 trillion (+29%). Furthermore, the share of the total VA (in constant prices of 2009) in the EU that is embodied by EU exports to the rest of the world has been steadily increasing between 1995 and 2011 and has reached in 2011 the level of 15%, compared to 10% in 1995.

This increase was predominantly driven by France (EUR +0.37 trillion), the United Kingdom (EUR +0.37 trillion), Spain (EUR +0.32 trillion) and Germany (EUR +0.31 trillion). They represented, respectively, 15.3%, 15.2%, 13.4% and 12.8% of the total increase. In 2011, 20% of the EU VA was generated in five largest member states: Germany, France, the United Kingdom, Italy and Spain.

In relative terms, during this period, a number of EU countries have shown growth in real VA by more than 75%, most of which are the new member states: Ireland[b] (+107%), Latvia (+91%), Estonia (+91%), Lithuania (+88%), Slovakia (+83%) and Poland (+79%). The largest member states exhibited much slower real growth rates: Italy (+11%), Germany (+17%) and France (+27%). So it can be observed that the more recent EU member states' contribution to EU trade has been increasing relative to the shares of the older and larger members.

[b]The *Irish Times* published an article on 12 July 2016 about the incredible GDP growth of Ireland: http://www.irishtimes.com/business/economy/handful-of-multinationals-behind-26-3-growth-in-gdp-1.2719047.

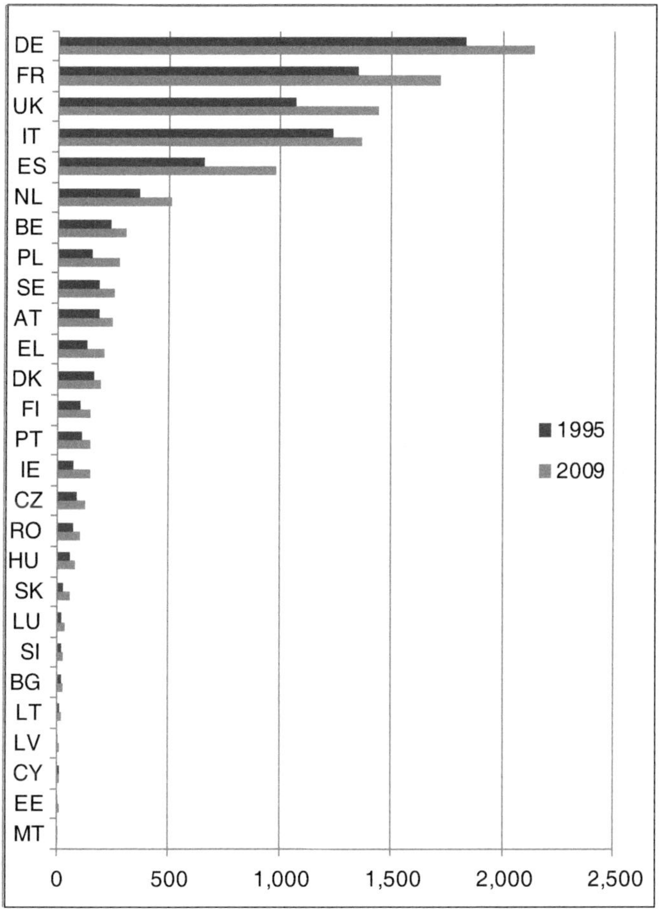

Figure 11.2 Value added at constant prices in 1995 and 2009 (billion EUR 2009 constant prices)[c] [13].

As the following discussion will be predominantly based on comparing the shares of different components of the income flows embodied in trade, the data on nominal VA in basic prices will be used, which is available until the more recent year 2011.

---

[c]Here and in figures later AT = Austria, BE = Belgium, BG = Bulgaria, CY = Cyprus, CZ = the Czech Republic, DE = Germany, DK = Denmark, EE = Estonia, EL = Greece, ES = Spain, FI = Finland, FR = France, HU = Hungary, IE = Ireland, IT = Italy, LT = Lithuania, LU = Luxembourg, LV = Latvia, MT = Malta, NL = Netherlands, PL = Poland, PT = Portugal, RO = Romania, SE = Sweden, SI = Slovenia, SK = Slovakia and UK = United Kingdom.

11.2.2 Embodied Value Added in EU Exports and Spillover Effects

In 2011, 83% of the total VA driven by EU exports was generated in the EU, while 17% was generated abroad (as shown in Fig. 11.3). As can be seen in Fig. 11.4 (red line), the share of the domestic EU VA generated by exports to the rest of the world has decreased compared to the 91% registered in 1995. Thus, the contribution to EU exports of VA created abroad has almost doubled, providing evidence of further internationalisation of global value chains, outsourcing labour, and capital consumption offshore.

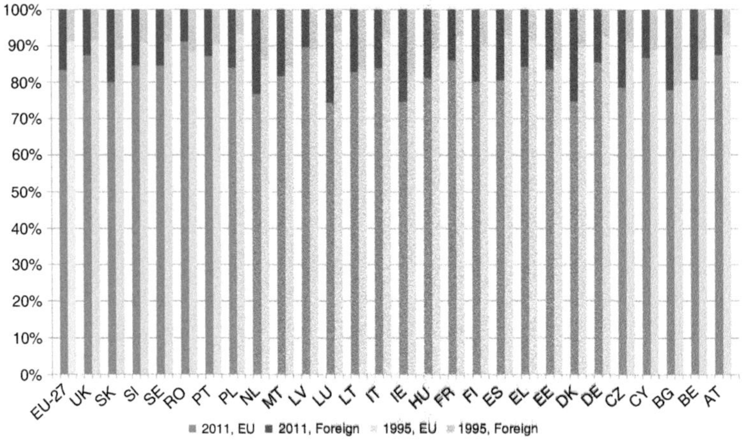

Figure 11.3 Total (EU and foreign) value added (in %) to exports of each member state and EU27 to the rest of the world for 1995–2011.

The corresponding share of foreign VA embedded in exports has increased in all member states. The strongest increases in foreign VA embodied in a country's exports were observed in Luxembourg (by 19%), Denmark (by 15%), Spain and the Czech Republic (by 12%). In several member states, the share of foreign VA has declined, though slightly: in Romania (by 3%) and in Latvia (by 1%).

When looking at the VA flows within the EU, we observe several distinctive patterns. Figure 11.5 provides a schematic overview of intra-EU income spillover effects underlying EU exports to the rest of the world. Each coloured line in the graphs (e.g., the dark-blue line linking Poland and Germany in 2011) represents the VA generated in Poland embodied in the exports of Germany to non-EU countries. The thickness represents the volume of the trade flows.

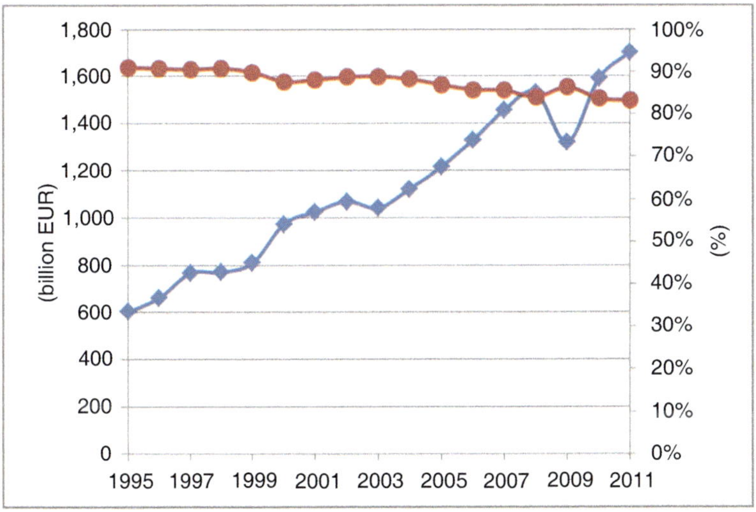

Figure 11.4 The domestic EU VA generated by EU exports in levels (blue) and as a percentage of the total EU VA (red) for 1995–2011.

Although the number and structure of the linkages are apparently stable (note that we have focused on the top 60[d] trade flows among EU member states), Fig. 11.5 shows that the hubs around the larger member states, like Germany, the United Kingdom and France, remained strong, as can be seen from the dense income flow linkages. It is interesting to identify that some particular income flow relationships have strengthened, for example, the increase in the VA in Belgium embodied in other member states' exports outside the EU or the increase in income in Poland that depends on the exports from Germany to non-EU countries.

A closer look at Europe's larger member states reveals that in the course of the last decades the exports of these countries became more reliant on the VA produced by the other members. Such a tendency is clearly visible in the case of Germany, France and Italy. On the other hand the United Kingdom's exports outside the EU

[d]Arbitrarily chosen as approximately twice the number of member states.

exhibited an opposite evolution and generated less VA in other EU member states (thinner coloured lines leaving from the UK arch in the direction of other countries) than other EU member states in the United Kingdom, due to their exports to the rest of the world (purple lines leaving from countries such as Germany, Ireland and Luxembourg and ending in the UK arch). In particular, the net difference is of EUR +28 billion in favour of the United Kingdom, while in other countries, such as Germany and France, the figures are negative (EUR –7 and –4 billion, respectively).

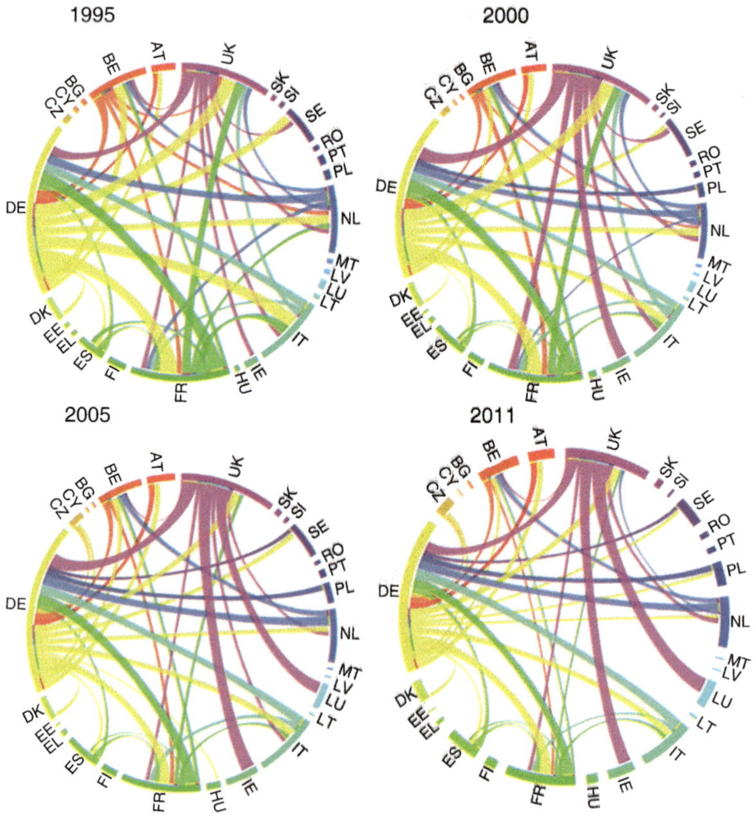

Figure 11.5 Map of intra-EU income spillover effects underlying EU exports to the rest of the world. *Note*: The line's thickness represents the volume of VA flowing from one country (same colour) to the other for the production of their exports outside the EU.

11.2.3 Results by Industry Type

Figure 11.6 shows that at the industry level, in 2011 the EU exports of machinery and transport equipment comprised, by far, the largest part of EU's exports. This export flow has, in its turn, generated EUR 5 billion of VA in the chemical industry; EUR 10 billion of VA in energy and EUR 6 billion of VA in the wood, paper and printing sector across the EU. On the services side, the sector of transport, trade and business services was responsible for the largest part of the EU services exports and generated an additional EUR 6 billion in machinery and transport equipment manufacturing and EUR 7 billion in the energy sector.

When looking at the foreign VA impact of the EU trade with the rest of the world, the general pattern is similar to that of the domestic effects.

In 2011, the machinery and transport equipment industries were responsible for 33.5% of the VA generated outside the EU and embodied in EU exports to the rest of the world. Exports of the transport, trade, and business services industries contributed to 24.1% of the total foreign VA in EU exports to the rest of the world; the chemical industries contributed 10.5%, the energy industry 10.8% and the other non-metallic and basic metals industries 9.5%. In most member states, exports to the rest of the world of the machinery and transport equipment industries and of the transport, trade and business services industries provided the largest contributions to foreign VA in EU exports.

It is also worth mentioning several examples of distinctive differences from the average EU picture. Two countries, Luxembourg and Cyprus, exhibited a different pattern, with by far the largest part of foreign VA being generated in business services (which also holds for EU VA; see Ref. [13]). Spain's and Portugal's exports generate a larger share of foreign VA in the energy sector. The share of foreign VA in the manufacturing exports of Slovakia and the Czech Republic is much larger than the European average (e.g., the car-manufacturing industry).

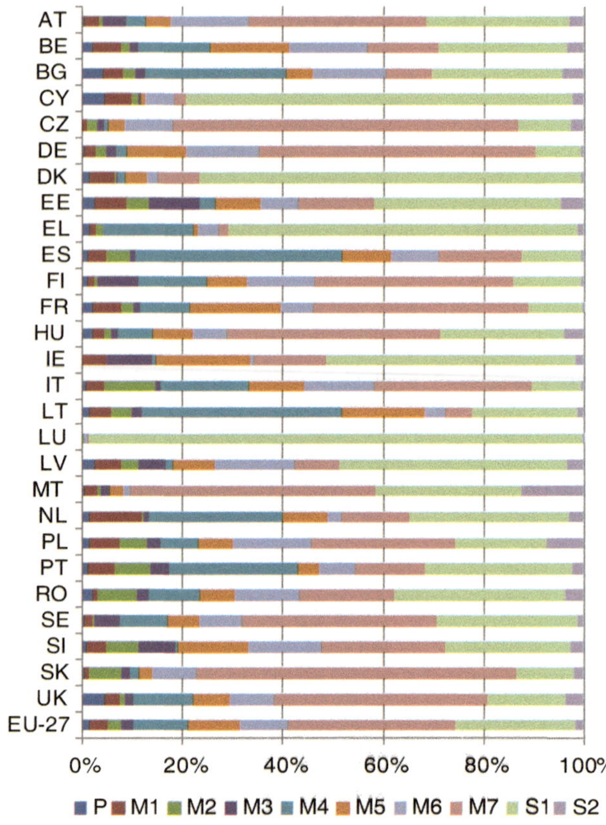

Figure 11.6 Foreign value added (in %) in exports of each member state to the rest of the world, by the exporting industry, for 2011. *Note*: P: primary; M1: food, beverages and tobacco; M2: textiles; M3: wood, paper and printing; M4: energy; M5: chemicals; M6: other non-metallic and basic metals; M7: machinery and transport equipment; S1: transport, trade and business services; and S2: other services [13].

11.2.4 Results by Trading Partner

There are several interesting observations that can be made when looking at the flows of EU VA generated by individual member states and the corresponding trading destination partners presented in Table 11.1. In 2011, 35% of the EU VA embodied in EU exports to the rest of the world was destined to the United States (16%), China (11%), Russia (5%) and Turkey (3%). In other words, 16% of the

EUR 1703 billion of the EU VA embodied in all EU exports was linked to trade with the United States. It is remarkable that this figure almost doubled for Ireland.

Table 11.1 EU value added generated in member states due to their EU exports to various trading partners (2011; % and billion EUR)[e]

	Exports to														
Value added in	AU	BR	CA	CN	ID	IN	JP	KR	MX	RU	TR	TW	US	RW	Total
AT	1%	4%	2%	13%	1%	1%	2%	2%	1%	4%	3%	1%	12%	53%	52
BE	2%	4%	3%	12%	1%	5%	3%	2%	1%	4%	4%	1%	18%	41%	59
BG	1%	1%	1%	5%	1%	0%	1%	1%	1%	11%	18%	0%	6%	54%	6
CY	1%	0%	0%	4%	0%	0%	0%	1%	0%	8%	3%	0%	3%	79%	2
CZ	2%	2%	3%	8%	0%	1%	2%	2%	1%	11%	4%	1%	14%	48%	23
DE	2%	3%	2%	14%	1%	1%	3%	2%	2%	5%	3%	1%	15%	45%	448
DK	2%	2%	2%	8%	1%	1%	3%	2%	1%	3%	1%	1%	13%	61%	37
EE	1%	2%	1%	6%	0%	1%	3%	1%	0%	19%	1%	0%	9%	54%	3
EL	0%	1%	1%	2%	0%	0%	0%	1%	1%	2%	4%	1%	5%	82%	18
ES	1%	3%	2%	5%	0%	1%	2%	1%	3%	4%	6%	1%	20%	50%	86
FI	2%	3%	2%	18%	1%	3%	4%	3%	1%	8%	2%	1%	14%	39%	29
FR	2%	4%	3%	11%	1%	1%	3%	3%	1%	4%	3%	1%	15%	49%	184
HU	1%	2%	1%	8%	1%	1%	2%	6%	1%	7%	4%	1%	9%	56%	20
IE	2%	0%	3%	4%	0%	0%	2%	1%	1%	1%	0%	1%	31%	53%	55
IT	2%	4%	2%	11%	1%	2%	3%	2%	2%	7%	5%	1%	14%	46%	171
LT	1%	0%	1%	2%	0%	0%	1%	1%	0%	20%	2%	0%	4%	68%	6
LU	0%	0%	1%	3%	0%	0%	0%	0%	0%	1%	1%	0%	2%	90%	16
LV	0%	0%	1%	4%	0%	1%	1%	0%	0%	12%	1%	0%	4%	75%	4
MT	1%	4%	1%	12%	0%	0%	7%	17%	3%	2%	2%	0%	12%	38%	1
NL	3%	2%	2%	15%	1%	1%	2%	2%	1%	4%	2%	2%	18%	43%	93
PL	1%	2%	3%	7%	0%	1%	1%	1%	1%	15%	5%	1%	11%	49%	43
PT	1%	10%	3%	5%	0%	1%	2%	1%	2%	4%	3%	0%	17%	51%	12
RO	1%	1%	3%	4%	1%	1%	2%	1%	1%	7%	10%	1%	6%	63%	16
SE	2%	2%	2%	11%	1%	1%	2%	2%	1%	4%	1%	1%	14%	55%	74
SI	1%	1%	2%	6%	0%	1%	1%	2%	1%	11%	4%	1%	9%	61%	5
SK	2%	2%	3%	12%	0%	1%	2%	1%	1%	20%	5%	1%	10%	41%	9
UK	3%	2%	5%	7%	0%	2%	3%	2%	1%	3%	3%	1%	23%	45%	231
EU-27	2%	3%	3%	11%	1%	1%	3%	2%	1%	5%	3%	1%	16%	48%	1,703

Source: [13].

More than 20% of the VA in EU exports to the rest of the world from Ireland, the United Kingdom and Spain ends up in goods and services sold to the United States. More than 15% of the contribution

[e]Here and in figures later AU = Australia, BR = Brazil, CA = Canada, CN = China, ID = Indonesia, IN = India, JP = Japan, KR = Korea, MX = Mexico, RU = Russia, TR = Turkey, TW = Taiwan, US = United States and RW = rest of the world.

to VA in EU exports to the rest of the world from Finland and the Netherlands was bound to China. EU exports to Russia generated more than 15% of the VA from Slovakia, Lithuania, Estonia and Poland, embodied in the EU exports to the rest of the world.

At the general economic sectors level, Arto et al. [13] demonstrate that in 2011, the largest share of the VA generated outside the EU (46%) embodied in EU exports to the rest of the world was generated in the services sector (same as in 1995). The manufacturing sector provided 29% (down from 37% in 1995), and 25% was produced in the primary sector (up from 17% in 1995). More than 50% of the VA from Japan and China, embodied in EU exports to the rest of the world, came from manufacturing. In other countries (with the exception of Russia) most of the VA to EU exports to the rest of world was generated in the services sector. In Russia most of the VA to EU exports to the rest of the world was generated in the primary sector (41%).

The relative shares of the aggregated sectors remained stable between 1995 and 2011 in most of the trade partner countries except for Turkey, India and, to a smaller degree, the United States, which exhibited a visible increase in the share of manufacturing VA embodied in the EU exports.

And at a more detailed sector classification level in Fig. 11.7 it is observed that in 2011, 39.7% of the VA generated outside the EU embodied in EU exports to the rest of the world was generated in the transport, trade and business services industries. The second-largest share (24.6%) was in the primary industries, followed by 9.2% in the machinery and transport equipment industries, 7.6% in the other manufacturing of non-metallic and basic metals industries and 6.7% in other services industries.

It is notable that the share of the primary sector in the VA generated outside the EU through the EU exports to its main trade partners is significantly higher than that of the manufacturing sectors. At the same time the large weight of the services sector's VA is evident and confirms the picture observed in the analysis of the impact of trade on employment outside the EU observed by Rueda-Cantuche and Sousa [16].

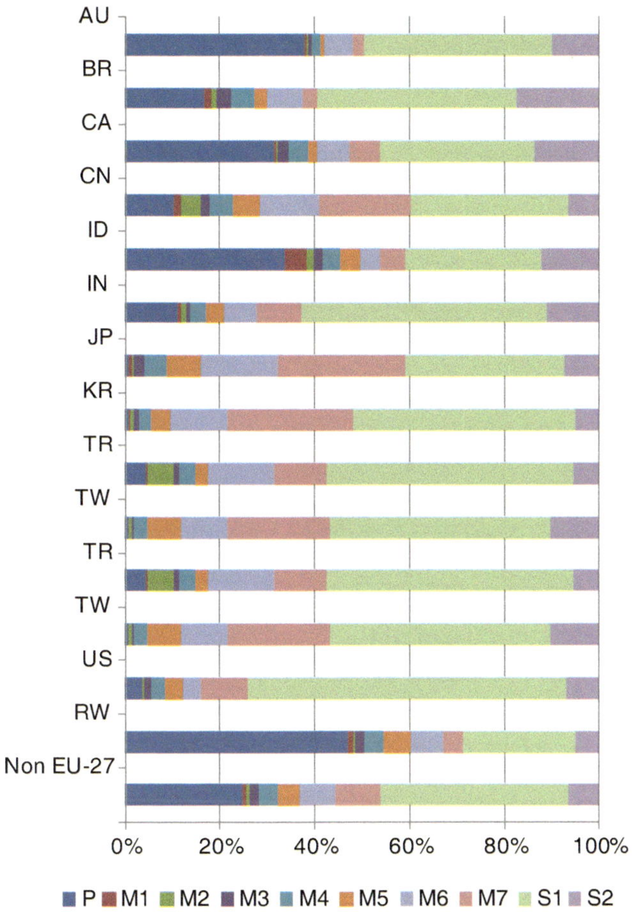

Figure 11.7 Value added by trading partner and industry in EU exports (in %) for 2011. *Note*: P: primary; M1: food, beverages and tobacco; M2: textiles; M3: wood, paper and printing; M4: energy; M5: chemicals; M6: other non-metallic and basic metals; M7: machinery and transport equipment; S1: transport, trade and business services; and S2: other services [13].

11.3 Conclusion

In the past decades expanding EU trade with the rest of the world had its impact not only on the structure of income flows within the EU but also on incomes of its trade partners. The total VA embodied in EU trade has been steadily increasing. This process has also been accompanied by the increasing share of its foreign VA component in the total trade support income.

The interdependence of the VA flows among EU members has been growing, and in particular, one could observe growing integration of new member states into the European single market's production activities driven by EU trade [17].

At the sectoral level, business services and machinery and transport equipment manufacturing are the main drivers behind EU trade with the rest of the world. Their contribution to EU exports to the rest of the world also comes together with additional VA creation in chemicals and energy sectors that can be viewed as the principal supporting activities.

In addition, we examined additional income flows driven by the expansion of EU trade that occurs beyond European borders. EU trade with the rest of the world mostly generates income in the services sectors of its trade partners. Nonetheless, it has been observed that the role of the primary sector in trade-driven foreign VA creation is also significant.

Among Europe's trade partners, the United States, China, Russia and Turkey are the main beneficiaries in terms of VA embodied in EU exports. Together they cover a quarter of all foreign VA embodied in EU trade and for some member states the share of an individual trade partner can reach as much as 30%.

The results of this study allowed us to take a closer look at the global value chains underlying EU trade dynamics. In general, the observed patterns highlight the ever-growing contribution of EU trade to the income flows in global value chains, where the degree of interdependence has been steadily increasing not only within EU but also with the rest of the world economy.

The study provides evidence that international trade has an ever-growing impact for generating income that benefits the citizens in all member states. These include economic activities that are selling directly to markets outside the EU and those that constitute the activities of suppliers of inputs to exporters that may be located in their own member state or elsewhere across the EU. Underlying this is a single market that is clearly an important pillar of EU competitiveness in the global marketplace as well as well-being of the EU population.

In addition, as global value chains expand, much income that is generated with support of EU exports benefits the producers and consumers beyond European borders. In global production networks, the trade relationship with a given partner must be assessed not only from the perspective of the size and dynamism of its market (for selling the final exports) but also from the standpoint of its importance as an input supplier (providing VA to the goods exported), because both aspects of the bilateral trade relationship have implications for income generation and well-being in the EU and elsewhere.

References

1. European Commission. (2015). *Trade for All: Towards a More Responsible Trade and Investment Policy* (Publications Office of the European Union, Luxembourg).
2. Sturgeon, T. J. and van Biesebroeck, J. (2011). Global value chains in the automotive industry: an enhanced role for developing countries?, *International Journal of Technological Learning Innovation and Development*, **4**, 181–205.
3. Fernández-Stark, K., Frederick, S. and Gereffi, G. (2011). *The Apparel Global Value Chain: Economic Upgrading and Workforce Development* (Center on Globalization, Governance & Competitiveness, Duke University, Durham, USA).
4. De Backer, K. and Miroudot, S. (2013). *Mapping Global Value Chains*, OECD trade policy paper no. 159 (OECD, France).
5. Boarini, R., Murtin, F. and Schreyer, P. (2015). *Inclusive Growth: the OECD Measurement Framework*, OECD statistics working papers, 2015/06 (OECD, France).
6. Daudin, G., Rifflart, C. and Schweisguth D. (2011). Who produces for whom in the world economy?, *Canadian Journal of Economics*, **44**, 1403–1437.
7. Hummels, D., Ishii, J. and Yi, K.-M. (2001). The nature and growth of vertical specialisation in world trade, *Journal of International Economics*, **54**, 75–96.
8. Johnson, R. C. and Noguera, G. (2012). Acconting for intermediates: production sharing and trade in value added, *Journal of International Economics*, **86**, 224–236.

9. Koopman, R., Powers, W., Wang, Z. and Wei, S.-J. (2010). *Give Credit Where Credit Is Due: Tracing Value Added in Global Production Chains*, NBER working paper no. 16426.
10. Trefler, D. and Zhu S. (2010). The structure of factor content predictions, *The Journal of International Economics*, **82**, 195–207.
11. Stehrer, R. (2012). *Trade in Value Added and the Valued Added in Trade*, WIIW working paper no. 81, Vienna.
12. Timmer, M. P., Dietzenbacher, E., Los, B., Stehrer, R. and de Vries, G. J. (2015). An illustrated user guide to the World Input–Output Database: the case of global automotive production, *Review of International Economics*, **23**, 575–605.
13. Arto, I., Rueda-Cantuche, J. M., Amores, A. F., Dietzenbacher, E., Sousa, N., Montinari, L. and Markandya, A. (2015). *EU Exports to the World: Effects on Employment and Income* (Publications Office of the European Union, Luxembourg).
14. Miller, R. E. and Blair, P. D. (2009). *Input-Output Analysis: Foundations and Extensions*, 2nd ed. (Cambridge University Press, New York).
15. Murray, J. and Lenzen, M. (2013). *The Sustainability Practitioner's Guide to Multi-Regional Input-Output Analysis* (Common Ground, Champaign, IL).
16. Rueda-Cantuche, J. M. and Sousa, N. (2016). EU exports to the world: overview of effects on employment and income, *Chief Economist Note* (European Commission, DG TRADE 2016).
17. Rueda-Cantuche, J. M., Sousa, N. Andreoni, V. and Arto, I. (2013). The single market as an engine for employment through external trade, *Journal of Common Market Studies*, **51**, 931–947.

Chapter 12

Global Trade, Pollution and Mortality

Christina Prell,[a] Klaus Hubacek,[a,b] Laixiang Sun[a] and Kuishuang Feng[a]
[a]*Geographical Sciences, University of Maryland, College Park, MD 20742, USA*
[b]*Department of Environmental Studies, Masaryk University, Brno, Czech Republic*
cprell@umd.edu

12.1 Introduction

Economic globalisation refers to flows of trade and capital among and between countries. A critical view on economic globalisation is that the historical forces influencing how and when a given country becomes integrated into global trade (and the world economy as a whole) condition the potential paths of development open to that country (e.g., Ref. [1]). Here, trade and other relations between countries act as structural mechanisms enabling wealthier, more core countries to maintain favourable terms of trade, which, in turn, negatively impact less developed, more peripheral ones in a variety of ways [2, 3].

In this chapter, our primary interest is in evaluating how countries' levels of integration in international trade give rise to

The Social Effects of Global Trade
Edited by Joy Murray, Arunima Malik, and Arne Geschke

ISBN 978-981-4774-55-0 (Hardcover), 978-020-3728-71-0 (eBook)
www.panstanford.com

inequality between countries in terms of emissions, wealth and mortality. In looking at these emissions inequalities, we make use of multi-regional input-output (MRIO) analysis to distinguish between emissions produced within a country via that country's economic activities (referred to as *production-based emissions*) and emissions that are triggered by a country's purchase of commodities and then allocated to that country in which purchases are made (referred to as *consumption-based emissions*). In this way, we are able to better capture how consumers trigger unequal distributions of production-based emissions. Finally, we are interested in how emissions and wealth work together to affect countries' mortality rates. Whereas production-based emissions are often experienced as a burden that local populations must endure in exchange for participation in global trade [4], economic growth and wealth are often seen as the main potential benefits of such participation [5, 6]. These trade-offs come together differently for different countries, such that the potential impacts of emissions experienced by local populations may be buffered by the benefits associated with more wealth. We consider whether such wealth indeed translates into a buffer against emissions' impacts by considering how countries' shares of global production-based emissions stand in relation to their shares of global value added (VA).

12.2 Global Trade, Emissions and the Effects on Mortality

International trade can be described as a system of interdependent economic relations that gives rise to structural features that shape characteristics and outcomes for countries. We adopt a network approach to studying international trade relations [6–8]. An important distinction is made between countries that are positioned in the network *core* versus its *periphery*. In network terminology, a core-periphery structure refers to a two-class partitioning where the core consists of a set of actors (or nodes) that are densely connected to one another and central to the entire network, that is, they form a well-integrated block and share a similar set of ties to others in the network. In contrast, the periphery refers to a class of countries that are more or less isolated from one another and linked to the rest of the network mainly via ties to the core.

Within this core-periphery structure, countries situated within the core are seen as being more integrated into the overall global trade network, consequently accruing larger benefits in the form of economic growth and/or development [5, 6] versus higher levels of emissions and resource exploitation in these less developed, more peripheral nations [7–9].

At the same time, more integrated, richer countries are seen as consuming a larger global portion of goods and services and thus triggering a disproportionately higher share of emissions through their consumption, the environmental impacts of which are to a larger extent outsourced to peripheral countries. As such, they are seen as being more accountable for a larger share of global emissions. These emissions along the entire supply chain allocated to the final consumer are captured in the consumption-based approach.

The questions that emerge are the following: Are core countries generating most of the emissions? Or have they managed to decrease levels of production-based emissions through externalisation, thus prompting considerable amounts of emissions in other countries?

In addition, we are concerned with identifying who is most affected by these emissions and how a country's level of integration might help mitigate the impacts of this pollution. As core countries tend to accumulate more wealth than less integrated ones, the fungible nature of this wealth should help societies adjust and potentially create a buffer against the harmful effects of emissions. Research has shown that the periphery suffers higher mortality rates than the core, and such high mortality rates are generally seen as resulting from the multiple structural disadvantages found within the periphery, namely weaker institutions and less environmental safeguards [8, 10, 11]. Taken together, we predict that more efficient countries would be the ones that would also hold lower mortality rates regardless of how much those countries might actually pollute.

12.3 Materials and Methods

Our trade data were extracted from the Eora database. Eora is an MRIO database that provides a time series of high-resolution input-output (IO) tables with matching environmental and social satellite accounts for 186 countries [12]. We use the provided MRIO tables for 26 sectors for a 20-year time span (1990–2010).

Our SO_2 emission data are at the sector level and were also collected from the Eora database. Our child and infant mortality data (each per 1000 live births) were downloaded from the United Nations Children's Fund's (UNICEF's) Millennium Development Goals database (visit http://mdgs.un.org/unsd/mdg/ Data.aspx for details). To handle skewness, we took the natural log of these mortality data before conducting any analyses. Using the square root or in this case the natural logarithm normalises the data distribution, which enables us to continue to use parametric statistics.

Bringing together data for all our variables of interest resulted in a final data set containing 172 countries.

To calculate a country's level of integration, we transformed the Eora data set following guidelines established by previous scholars [6, 13]: First, we aggregated all 26 sectors to form one country-by-country trade matrix consisting of value data. A single, value trade network data set is often preferred for operationalising ideas of coreness or integration, as a country's position not only is determined through the quantity and patterning of trading ties to others but also considers the trade volume of those ties [13]. In addition, as rows in the trade matrix correspond to exports and columns to imports, we took the trade matrix and its transpose and then summed the two together to arrive at a symmetrised matrix that combines, for each country, information on that country's exports and imports. By summing the export matrix to its transpose (i.e., the import matrix), we focused attention on the structure of trade, as opposed to the directionality of trade ties [6, 13].

To measure integration, we made use of the continuous coreness procedure [14] for measuring countries' levels of integration within the world trade network. This procedure fits a core-periphery model to an observed network to identify the extent to which the observed network approaches an ideal core-periphery structure. In an ideal core-periphery structure, the core is a block of actors who are tied to one another and, in addition, have ties with many other actors in the network. To be core, then, is to be highly central in one's own right, as well as part of a dense block of other highly central actors. In contrast, peripheral actors form a second block, in which members are largely isolated from one another and any ties they do hold are with the core. The coreness procedure [14] proceeds in determining what type of partitioning of actors (in our case, countries) in the observed network most closely brings that network towards an ideal core-periphery partitioning. The procedure results in a vector of scores assigned to countries, which range between 0 and 1, the higher values indicating a country being more core and lower values indicating the country being more peripheral. The advantage of this procedure is that it results in a ratio-scaled vector of scores, thus enabling a higher degree of precision in making cross-country comparisons. We refer to this resulting vector of scores as our level of integration measure for countries.

To measure a country's gains in terms of VA in relation to its environmental costs or losses along global supply chains, we developed a global pollution-wealth ratio (PWR) by dividing a country's share of global SO_2 emissions ($\%SO_2$) by its global share of VA (%VA), such that ($PWR = \%SO_2/\%VA$). This enables us to see how countries compare in terms of global shares of emissions and wealth.

Our major control variables include population size (in 1,000,000s) and urbanisation (percentage of population estimated to live in urban areas in a country), both taken from the World Bank's database (http://data.worldbank.org/indicator/SP.URB.TOTL.IN.ZS). Research has shown that population size is positively linked to forms of environmental degradation, including air pollution (e.g., see Refs. [15, 16]). Similarly, research has shown a positive link between a country's level of urbanisation and environmental degradation (e.g., see Refs. [16, 17]). Urbanisation has also been shown to be positively linked to infant mortality in the periphery [11], although in the core, urban centres have historically been places where wealth and other key resources necessary for a higher quality of life become concentrated (e.g., see Ref. [18]), thus potentially mitigating the health threats of air pollution. In addition, we have included countries' health expenditures as percentages of GDPs and countries' fertility rates as control variables for regression models predicting infant and child mortality. Data for both these variables were downloaded from the World Bank (http://www.worldbank.org).

For assessing which countries are causing SO_2 pollution, we regressed countries' integration scores against our two SO_2 pollution measures (consumption based versus production based), controlling for population size and urbanisation. Finally, to gauge the extent to which countries suffer from their pollution costs, we regressed mortality rates on levels of integration, the PWR and production-based SO_2, alongside other controls.

12.4 Results

We start with results for production-based SO_2 and consumption-based SO_2, as shown in Table 12.1.

Here, we see that for both emission outcome measures, integration holds a positive, highly significant coefficient, even after the controls are entered into the model. This finding suggests that the levels of countries' SO_2 pollution are positively associated with their levels of integration in global trade. In addition, consumption-based SO_2 is more strongly correlated to integration than production-based SO_2. In particular, models 2 and 4 show that the integration coefficient (elasticity) is much stronger (i.e., more than doubled) in relation to consumption-based SO_2 than production-based SO_2, once we control for population size and urbanisation. Finally, the explanatory power of the models increases once the control variables are introduced, as reflected in the R^2 values. As such, we see evidence that core, well-integrated countries are the larger emitters of SO_2 emissions via manufacturing and also appear to be the main externalisers of emissions.

To what extent does an increase in the PWR have a pay-off, in terms of mortality rates, for individual countries, or in other words, would lower PWR scores translate into an economic buffer to help reduce mortality rates for infants and children? Table 12.2 shows regression model results exploring this question.

Table 12.1 Pollution regressed on countries' levels of integration

	Production-based SO_2 (ln)		**Consumption-based SO_2 (ln)**	
	Model 1	**Model 2**	**Model 3**	**Model 4**
	b	*b*	*b*	*b*
Integration (ln)	1.067***(0.249)	1.175***(0.215)	2.673***(0.192)	3.409***(0.215)
Population (ln)		0.904***(.015)		0.670***(0.023)
Urbanisation		0.032***(0.001)		0.022***(0.001)
Constant	6.285***(0 .690)	4.428***(0.616)	11.035***(0.531)	11.305***(0.615)
Fixed?	Year, country	Year#	Year, country	Year#
Observations	3,486	3,483	3,591	3,588
Wald X^2	36,837.92	87,103.48	11,710.32	238,155.10
R squared	0.965	0.815	0.97	0.866

*$p < 0.10$, **$p < 0.05$, and ***$p < 0.01$. These are unstandardized *b* values. Standard errors in parentheses.
#The introduction of the country-fixed effect led to a highly singular variance matrix, implying high collinearity.

Table 12.2 Integration predicting infant and child mortality under the age of 5 years

	Infant mortality per 1000 (ln)		Child mortality per 1000 (ln)	
	Model 5	**Model 6**	**Model 7**	**Model 8**
Integration (ln)	-.0003(.024)	–.394**(.131)	.001(.026)	–.5481***(0.155)
PWR (ln)		.0501***(.012)		.072***(.0116)
Product SO_2 (ln)		.017(.012)		.011(.013)
Fertility (ln)		1.242***(.054)		1.117***(.051)
Health in %GDP		–.034***(.006)		–.044***(.007)
Urbanisation		–.016***(.001)		–.014***(.001)
Constant	5.291***(.071)	2.058***(.401)	4.912***(.074)	1.452**(.474)
Fixed?	Year, country	Year#	Year, country	Year#
N or observations	3,591	2,567	3,591	2,567
Wald X^2	721,148.80	23,563.60	640,638.06	24,370.93
R squared	0.989	0.925	0.9895	0.936

*$p < 0.10$, **$p < 0.05$, and ***$p < 0.01$. These are unstandardized coefficients. Standard errors in parentheses.

#The introduction of country-fixed effect led to a highly singular variance matrix, implying high collinearity.

For both sets of models in Table 12.2, we see that the countries' levels of integration hold a negative and statistically significant relationship with mortality rates, suggesting that more core countries have lower mortality rates. In addition, countries' PWR scores have a positive and statistically significant relationship to mortality, suggesting that countries with higher shares of emissions in relation to shares of VA suffer higher rates of mortality. In short, it appears that more core countries with higher economic buffers experience lower mortality rates.

12.5 Discussion and Conclusion

Taken together, when we ask ourselves who is most affected by SO_2 emissions, our findings show that more integrated countries experience lower mortality rates, yet they appear to do so via a number of mechanisms. One such mechanism is a country's PWR score. Low PWR scores (implying greater efficiency) appear to help reduce mortality, and thus, it appears that higher levels of integration help countries acquire stronger economic buffers to mitigate the negative impacts of SO_2 on mortality. In addition, higher levels of integration coincide with higher levels of urbanisation, higher health expenditures and lower fertility rates. All these mechanisms thus appear to be working together to give more core, integrated countries an advantage over less core ones with regard to mortality rates. Thus, the simple answer to the question 'Who is most affected by pollution emissions?' is that less core countries are the ones most affected.

In terms of our contribution to the literature, we have shown, quite clearly, evidence for well-integrated core countries being both major polluters as well as major externalisers of emissions. Further, our focus on SO_2 has enabled us to see how global trade patterns can have real local impacts in the form of mortality rates. As such, we have moved beyond fundamental concerns pertaining to inequalities resulting from globalisation (e.g., wealth and emissions) to show how such inequalities translate into real life-or-death issues for given societies.

References

1. Chase-Dunn, C. K. (1998). *Global Formation: Structures of the World-Economy*, updated ed. (Rowman & Littlefield, Lanham).
2. Arrighi, G. and Drangel, J. (1986). The stratification of the world-economy: an exploration of the semiperipheral zone, *Review*, **10**, 9–74.
3. Chase-Dunn, C., Kawano, Y. and Brewer, B. D. (2000). Trade globalization since 1795: waves of integration in the world-system, *American Sociological Review*, **65**, 77–95.
4. Copeland, B. R. and Taylor, M. S. (2004). Trade, growth, and the environment, *Journal of Economic Literature*, **42**, 7–71.
5. Mahutga, M. (2006). The persistence of structural inequality? A network analysis of international trade, 1965–2000, *Social Forces*, **84**, 1863–1889.
6. Clark, R. (2010). World-system mobility and economic growth, 1980–2000, *Social Forces*, **88**, 1123–1151.
7. Prell, C., Feng, K., Sun, L., Geores, M. and Hubacek, K. (2014). The economic gains and environmental losses of US consumption: a world-systems and input-output approach, *Social Forces*, **93**, 405–428.
8. Prell, C., Sun, L., Feng, K. and Myroniuk, T. W. (2015). Inequalities in global trade: a cross-country comparison of trade network position, economic wealth, pollution and mortality, *PLOS ONE*, **10**(12), e0144453.
9. Jorgenson, A. K. (2012). The sociology of ecologically unequal exchange and carbon dioxide emissions, 1960–2005, *Social Science Research*, **41**, 242–252.
10. Leonard, H. J. (2006). *Pollution and the Struggle for the World Product: Multinational Corporations, Environment, and International Comparative Advantage* (Cambridge University Press, Cambridge).
11. Jorgenson, A. K. and Rice, J. (2012). Urban slums and children's health in less-developed countries, *Journal of World-System Research*, **18**, 103–116.
12. Lenzen, M., Moran, D., Kanemoto, K. and Geschke, A. (2013). Building Eora: a global multi-regional input-output database at high country and sector resolution, *Economic Systems Research*, **25**, 20–49.
13. Prew, P. (2010). World-economy centrality and carbon dioxide emissions: a new look at the position in the capitalist world-system and environmental pollution, *Journal of World-Systems Research*, **16**, 162–191.

14. Borgatti, S. P. and Everett, M. G. (1999). Models of core/periphery structures, *Social Networks*, **21**, 375–395.

15. Dietz, T., Rosa, E. and York, R. (2007). Driving the human ecological footprint, *Frontiers in Ecology and the Environment*, **5**, 13–18.

16. Jorgenson, A. K. and Clark, B. (2012). Are the economy and the environment decoupling? A comparative international study, 1960–2005, *American Journal of Sociology*, **118**, 1–44.

17. Roberts, J. T. and Parks, B. C. (2007). *A Climate of Injustice: Global Inequality, North-South Politics, and Climate Policy* (MIT Press, Cambridge).

18. Thomas, A. R. (2012). Urbanization before cities: lessons for social theory from the evolution of cities, *Journal of World-System Research*, **18**, 211–235.

Part IV

Social Effects of Global Trade and Standard Business Practice

Chapter 13

Socially Sustainable Supply Chains

Graham Sinden
Climate Change and Sustainability Services, Ernst & Young,
200 George Street, Sydney, NSW, Australia
graham.sinden@au.ey.com

13.1 Introduction

In 1958, Leonard Read published his now-famous essay *I, Pencil*, a first-person description of the inputs and processes needed to make a pencil. In this essay, the pencil makes a bold claim, 'Not a single person on the face of this earth knows how to make me' [1], a reference to the specialisation and diversification of supply chains that draw intellectual, financial, human, manufacturing, social and natural capital [2] together to create goods and services.

Fast-forward half a century, and the underlying premise of supply chain complexity has only expanded. This expansion is not an objective in itself but reflects the opportunities that the increasing internationalisation of trade and production provides for producers and consumers. In this environment, the implications of the claim of *I, Pencil* extend beyond the ability of an individual to understand the production process, to the ability of producers and consumers

The Social Effects of Global Trade
Edited by Joy Murray, Arunima Malik, and Arne Geschke

ISBN 978-981-4774-55-0 (Hardcover), 978-020-3728-71-0 (eBook)
www.panstanford.com

to understand the social implications of their supply chain relationships, and the potential implications of these relationships on businesses and decision-making.

This creates challenges for businesses and for regulators. For businesses seeking to demonstrate their social or environmental credentials, the obfuscation of complex supply chain linkages can hinder their capacity to articulate superior performance; worse, it can create unseen risks for businesses that may unknowingly rely on inputs from sectors or other businesses that do not hold the same environmental or social values. For regulators, the complexity and internationalisation of supply chains may limit their ability to achieve policy outcomes through domestic regulatory activity alone. This is well-recognised in common issues such as climate change and air pollution, but is increasingly being recognised in the social sphere as an element of human rights.

Chapters 1 and 2 of this book set out the origins of global trade and the overarching framework for the regulation of human rights responsibilities in the supply chain, while Chapter 3 sets out analytical methods (in particular, multi-regional input-output [MRIO] models) to support supply chain analysis. This chapter builds on this foundation, exploring some aspects of the relationship between supply chain issues and businesses seeking to navigate supply chain challenges.

13.2 Regulatory Drivers of Human Rights and the Supply Chain

Over the past decade, human rights issues in the supply chain have received greater attention from regulators, which has follow-on consequences for business risk and decision-making.

While the focus of the UK Modern Slavery Act of 2015 is arguably the creation of sanctions relating to human trafficking and slave labour, Section 54 of the act establishes minimum requirements for disclosure 'of the steps the organisation has taken during the financial year to ensure that slavery and human trafficking is not taking place (i) in any of its *supply chains* . . .' [3]. By referencing supply chains in the legislation, policy-makers have sought to expand the application of this law beyond UK national boundaries: this recognises the role that the internationalisation of trade and production plays within modern economies. But this also creates challenges for businesses, particularly concerning the extent to which activities that may exist at the very periphery of the supply chain should be included, assuming that these relationships can be identified in the first place.

With the United Kingdom having historically embraced trade liberalisation, the UK Modern Slavery Act has garnered significant attention from both domestic and international businesses that are captured within its scope. But it is not the first, nor the only, policy initiative in this area. In 2010, the California Transparency in Supply Chains Act established a requirement for certain businesses operating in California to disclose their 'efforts to eradicate slavery and human trafficking from [their] direct supply chain'. An important distinction between the California and UK approaches is the direct qualifier to supply chains seen in the California legislation: this suggests a more limited view of supply chains than that taken in the drafting of the UK Act. The Dodd–Frank Wall Street Reform and Consumer Protection Act of 2010 also targets company supply chains; however, it takes a very different approach to the UK and California legislation. Rather than focusing on a specific issue such as slavery, the Dodd–Frank Act extends responsibility up the supply chain for users of a particular commodity, conflict minerals,[a] specifying both the commodity and the country of origin (the Democratic Republic of the Congo or an adjoining country) [4].

More recently, Australia commenced an inquiry into whether it should legislate to address modern slavery. With parliamentary comments framing this as 'an opportunity to explore whether Australia's laws can be improved to prevent modern slavery, both in Australia and in the supply chains of businesses and organisations which operate in Australia and overseas' [5], there is the prospect that legislation resulting from the inquiry would follow the United Kingdom's more expansive view of supply chain responsibility.

These examples suggest a direction of interest for regulators and, hence, an emerging risk and compliance issue for businesses. However, it is not solely regulation that is driving corporate consideration of social supply chain issues; some businesses are moving ahead of regulation, supported by emerging non-regulatory initiatives.

---

[a]Include tantalum, tin, gold and tungsten.

13.3 Non-Regulatory Initiatives for Social Supply Chain Linkages

While there are emerging examples of state and national legislatures addressing social issues within supply chains, this has to date had its focus limited to slavery and human trafficking. For businesses seeking to more broadly demonstrate their social credentials, non-government initiatives that frame more broadly the social issues arising in supply chains are emerging.

This is becoming increasingly relevant for businesses, with global investor surveys showing the increasing relevance of environmental, social and governance issues to mainstream investors in recent years. For example, in a survey of 320 investors representing banks, pension funds, insurers, and others, 92% agreed or strongly agreed that 'over the long term ESG issues . . . have real and quantifiable impacts' [6].

Drawing support and inspiration from processes such as the UN *Guiding Principles on Business and Human Rights*, social supply chain assessment frameworks have been developed to aid transparency and reporting. The Corporate Human Rights Benchmark (CHRB) is one such initiative seeking to standardise a method for assessing human rights impacts in the business environment [7]. The benchmark takes a broad view of the supply chain to include direct and indirect suppliers and sets out six measurement themes that address issues beyond slavery, such as payment of a living wage by suppliers to their workers, minimum protections for child labour, and health and safety. The Social Capital Protocol is another recent initiative that provides businesses with a method for quantifying supply chain impacts [8]. In contrast to the CHRB's benchmarking approach, the Social Capital Protocol is designed to be implemented by individual businesses, with the results supporting a range of functions from strategy and cost-benefit analysis to external reporting.

These and other non-governmental organisation (NGO) or private sector initiatives provide a framework for operationalising social supply chain measurement in a business setting.

13.4 Finance and Social Supply Chain Linkages

While the focus of social supply chain initiatives, whether regulatory or voluntary, tends to be on the physical relationships within supply chains, national and international capital flows are the mechanism through which production and trade are financed. As a result, there is a link between the capital flows that enable investment and production, and the potential for social supply chain issues to arise from those (or related) investments.

Socially responsible or ethical investing has a long track record, emerging first with boutique investment firms and then, increasingly, with mainstream investors offering ethical investment options alongside more traditional products. These typically apply a 'filter', either negative (investment in a security is allowed unless excluded) or positive (investment in a security is only allowed if it meets a pre-determined benchmark), to limit the scope of activities that are available for investment through the product.

Increasingly, this interaction between finance and non-financial drivers is being considered by key actors in the global financial system, with the focus of attention being the potential impacts of climate policy and climate change on investments. Initiatives such as the Financial Stability Board's Task Force on Climate-Related Financial Disclosures [9] and the United Nations Environment Program's Finance Initiative [10] are seeking to raise awareness and provide guidance to companies on the linkages between finance and climate.

In Australia and internationally, examples are emerging of banks assessing and disclosing climate metrics for their business. For example, the Commonwealth Bank of Australia has for the past two years published an analysis of the financed emissions arising from its lending portfolio [11]. In this particular approach, MRIO data are leveraged to provide the whole-of-economy coverage required for assessment of the bank's lending portfolio, supplemented with additional company-specific data. Other financial institutions have carried out different analyses, some of which also rely on extended MRIO models that can deliver supply chain greenhouse gas emissions data. And while these approaches have yet to be extended to social supply chain issues, increasing interest in this area may provide the impetus for developing such models.

13.5 Analytical Methods Supporting Social Supply Chain Analysis

A primary challenge for businesses seeking to understand, and potentially monitor and report on, social supply chain issues is the ability to quantify social aspects of their supply chains. However, supply chains for even apparently simple products (such as a pencil) may be highly complex, creating a data gathering burden on businesses and potentially limiting comparability between different analyses. Simplification of the analysis is one route through this complexity.

Limiting the scope of the supply chain, for example, to direct (or tier 1) relationships only, may be attractive as it reduces the complexity of the supply chain to be considered. But this simplicity comes at the expense of completeness, and in complex supply chains it may result in material social challenges not being identified due to their apparent distance from the organisation carrying out the analysis.

Another potential simplification is with the measurement metrics themselves. By focusing on slavery, the UK Modern Slavery Act has a narrow but clearly defined purpose. By contrast, the CHRB and the Social Capital Protocol take a far more expansive view of the potential social supply chain issues facing businesses. While this approach provides a comprehensive understanding of social supply chain issues of businesses, it demands additional resources that may act as a barrier to implementation.

The focus of this book on the extension of MRIO models into the social sphere offers businesses a pathway forward for understanding the social impacts and challenges in their supply chains, without the need to limit the analysis (e.g., to tier 1 relationships) to make the task manageable. A purely MRIO-based approach may support a 'heat map' analysis of the supply chain, identifying those sectors and countries where challenging social issues are more likely to arise for a business, and allowing further targeting of these. The case studies presented in Part III of this book use MRIO-based approaches to identify and analyse wide varieties of social issues along international supply chains.

Further evolution of MRIO approaches to social impacts may see the development of hybrid social supply chain models, leveraging the supply chain complexity represented by the MRIO model but tailoring the model to more closely reflect the specific supply chain relationships of individual businesses. Achieving this outcome would provide a powerful method for a business to understand its social supply chain challenges, not just presenting a 'social good' differentiator between itself and its competitors, but also meeting increasing regulatory oversight of key social supply chain issues and investor demands for quantifiable non-financial data related to its activities.

13.6 Conclusion and Outlook

A number of drivers are combining to drive corporate interest in social supply chain issues and to find new and meaningful ways to measure and report on them. Emerging regulatory drivers provide the 'stick' for some companies in some jurisdictions; however, increasingly investor demands for the disclosure of non-financial measures of performance, and the potential for companies to differentiate themselves from competitors, are also powerful drivers of action.

Leading analysis in the finance sector is already showing the relevance of analytical techniques such as MRIO to non-financial analysis and reporting. International initiatives to bring recognition of the importance of non-financial metrics and consistency to analysis and reporting techniques will further embed these types of analyses as standard business practices. The adaptation of MRIO techniques to social supply chain issues would bring further understanding of the linkage between businesses and their supply chains, supporting business, investor and consumer decision-making in an increasingly trade-oriented world.

Disclaimer

Any views expressed by Dr. Sinden are his own and not necessarily those of Ernst & Young.

References

1. Read, L. E. (1958). I, pencil, in *The Freeman* (Foundation for Economic Education, Irvington-on-Hudson, NY).
2. International Integrated Reporting Council. (2013). The International Integrated Reporting Framework, http://integratedreporting.org/resource/international-ir-framework/.
3. UK Government. (2015). Modern Slavery Act 2015.
4. US Securities and Exchange Commission. (2012). *SEC Adopts Rule for Disclosing Use of Conflict Minerals*, https://www.sec.gov/news/press-release/2012-2012-163htm#.Uxdchc5NDwU.
5. Crewther, C. J. (2017). Private members' speech on international development assistance, Canberra, Australia.
6. Ernst & Young (2017). Is your nonfinancial performance revealing the true value of your business to investors? http://www.ey.com/gl/en/services/assurance/climate-change-and-sustainability-services/ey-nonfinancial-performance-may-influence-investors.
7. Corporate Human Rights Benchmark. (2016). *CHRD Pilot Methodology*.
8. World Business Council for Sustainable Development. (2017). *Social Capital Protocol*, http://www.wbcsd.org/Clusters/Social-Impact/Social-Capital-Protocol.
9. TCFD. (2017). *Task Force on Climate-related Financial Disclosures*, https://www.fsb-tcfd.org/.
10. UNEP. (2017). *UNEP Finance Initiative*, http://www.unepfi.org/.
11. Commonwealth Bank of Australia. (2016). *Assessed Carbon Emissions of the Group's Business Lending Portfolio*, https://www.commbank.com.au/content/dam/commbank/about-us/docs/assessed-carbon-emissions-report-2016.pdf.

Index